JESSICA
BECK

The Nature of Natural History

The Nature of

Natural History

MARSTON BATES

With a new Preface by
HENRY S. HORN

PRINCETON UNIVERSITY PRESS
PRINCETON, NEW JERSEY

PUBLISHED BY PRINCETON UNIVERSITY PRESS,
41 WILLIAM STREET, PRINCETON, NEW JERSEY 08540

LIBRARY OF CONGRESS CATALOGING-IN-PUBLICATION DATA

BATES, MARSTON, 1906–1974.
THE NATURE OF NATURAL HISTORY/MARSTON BATES; WITH A NEW PREFACE BY
HENRY S. HORN.
P.   CM.—(PRINCETON SCIENCE LIBRARY)
REPRINT. ORIGINALLY PUBLISHED: NEW YORK: SCRIBNER, 1950.
INCLUDES BIBLIOGRAPHICAL REFERENCE.
ISBN 0-691-02446-4 (ALK. PAPER)
1. NATURAL HISTORY. 2. BIOLOGY. 3. SCIENCE. I. TITLE. II. SERIES.
QH45.B35   1990
574—DC20   90-8143
FIRST PRINCETON SCIENCE LIBRARY PRINTING, 1990

10 9 8 7 6 5 4 3 2 1

PRINTED IN THE UNITED STATES OF AMERICA BY
PRINCETON UNIVERSITY PRESS, PRINCETON, NEW JERSEY

THIS BOOK IS FOR MY MOTHER-IN-LAW
MARIAN HUBBARD BELL FAIRCHILD.
DAUGHTER, WIFE AND MOTHER OF SCIENTISTS,
SHE KNOWS WELL HOW THE SPECIES SHOULD BE
FED AND EXERCISED, AND FORGIVES ITS
FOIBLES AS SHE ESTEEMS ITS VIRTUES.

CONTENTS

Preface to the
Princeton Science Library Edition

SINCE Marston Bates wrote *The Nature of Natural History* in 1950, there have been intellectual revolutions in natural history and in the neighboring sciences. Continental drift has become established as a fact of earth's history, underscoring the importance of unique historical events in evolution. The fossil history of life on earth has been pushed much earlier, and many strange and wonderful creatures have come to light. Molecular biology has forged the mechanistic link between mutations and their physiological effects, and evolutionary biologists now routinely uncover the details of the origins of species, families, and orders of plants and animals by reading the code inherited in DNA.

Ecology has been conceptualized and mathematized. Although initially there was a chaotic period when theoreticians and field workers either deified or reviled each other, they now work together, and increasing numbers of scholars are equally experienced in field observation and technical analysis. Biogeography now has added some mechanistic underpinnings to its old historical empiricism. The field of behavioral ecology has arisen from analytical studies of the decisions that animals make, be they conscious or not.

These are exciting times for biologists, and particularly for devotees of natural history. But not all is well; the revolutions have increased the amount of specialized expertise needed to practice increasingly separate sciences. Ecologists, molecular biologists, and geologists are training themselves and their students for their own specialties. It is a wise and lucky university with a natural history museum and comparative physiology, embryology, and functional anatomy faculties—resources that Bates cites as crucial to any serious biologist's training. Furthermore, there is a terrible tendency for scientists who depend on expensive machinery and large grants to disparage the intellectual challenges of natural history, which can be done with only diligence, perception, imagination, and judgment.

So Bates's original goal is suddenly modern again. He defines and defends natural history as an intellectual enterprise, and sets it in the broad context of traditional biology. It is a tribute to his judgment and his style that nearly everything he said is still fresh, and very little of it has been overturned by the revolutions since 1950. I have added some updating notes and references to the end of the book. Nevertheless, most of Bates's original bibliography and his wise and engaging notes have withstood the test of time.

There are a few points where cautious reading is needed because of the excesses of intervening intellectual generations. These concern "purpose" in evolution (pp. 57, 210, 250–251), the good of the species or community as a criterion for natural selection (pp. 108, 154, 217, 223, 236), and occupational sexual stereotypy (pp. 252–263). In all these instances, I am sure that Bates, were he alive today, would write more cautiously.

I first read *The Nature of Natural History* in 1956, as a high school student earning pin money by topping up the alcohol in preserved specimens in the basement of Harvard's Museum of Comparative Zoology. I remember discussions of the book over "smokes" on steps outside the main entrance (Bates describes the setting on p. 42). Everyone who read the book was impressed with its substance, clarity, humor, eloquence, and timeliness. My most recent rereading is as a biology professor at Princeton in 1989. I'm still impressed on all counts.

—HENRY S. HORN

CHAPTER I

The Science of Natural History

THERE is a temptation to start a book with some grand phrase, some broad statement that will lead the reader on into the details of the text. The movie people often use such a device, starting with the camera aimed at an immensity of sky and clouds, lowering it to make a sweep across a wide landscape of forests and fields until one village is picked out, one street, one house. Within the house, the focus finally comes to rest on Dorothy, sitting quietly at her spinning wheel, her outward calm a cover for some seething turmoil of emotions. Only then does the story begin to develop.

If we used this same trick here, we could skip the incomprehensible universe of the astronomers and start by focussing on our planet, tiny by astronomical standards but big enough, in all conscience, to its inhabitants. Our first camera shot would be of this earth whirling endlessly on its axis, blindly following its elliptical course around the sun. We might stop to notice that its distance from the sun was just right to produce the temperatures that we find comfortable (as well as some uncomfortable ones), that its twistings and turnings gave us day and night and an annual succession of seasons and resulted in a curious zoning of the planetary surface from equatorial tropics to polar arctic.

Examining this surface more closely, we would see that most of it was covered with a thin film of water, that the dry parts were crinkled into corrugations of mountains in some places, spread out as broad plains in others. As we came down through the mats of cloud, we would make out details of lakes, rivers, bays, forests, savannahs—the endlessly variegated environmental conditions of the planetary surface.

If we carried the thing out in proper Hollywood style, we would land in the middle of a biological laboratory where there would be a man in a white jacket peering at a drop of water with a microscope. He would stand aside and allow us to look through the eyepiece at the world of subvisible creatures that he had been studying. There, certainly, we would see marvellous things: diatoms and desmids with their crystalline symmetry; bright green filaments of algae forming thickets through which jelly-like ciliates picked their leisurely way; bacterial rods trembling constantly from molecular bombardment, the ceaseless vibration called Brownian movement. Then, having settled our eye definitely on some particular bug, we could start to describe its life, its history, its relations with the other organisms surrounding it, and so gradually build up a picture of biological processes.

But such an approach would be misleading. My errand, in writing this book, is missionary; but it is not to impress with the facts of biology, or with the marvels of what we have learned about the processes of life. Rather, my errand is to arouse interest in an attitude, to explain a point of view, using facts only insofar as they may be necessary as illustrations for this attitude, this point of view.

The facts of the various sciences have been fairly well publicized; their impact has altered our material environment, affecting the course of our daily lives in all sorts of ways, many obvious, many subtle and indirect. The facts of biology have penetrated the nursery, the kitchen, the garden almost as much as they have the farm or the hospital. But the attitude of biology, the general method of science, seems to have made no corresponding contribution to our culture, to have resulted in no comparable alteration of our intellectual processes. Science remains a rather mysterious affair, cultivated by a special priesthood, guarded by an unintelligible jargon. We are all exposed to science in school: but the dissection of the frog or the smells of the chemical laboratory have about as much relation to the living discipline of science as do the rules for the ablative absolute to the living language in which Cicero played his games of politics.

It is a difficult assignment, to convey an attitude. I do not think it can be done by describing facts, or by arranging facts into narrative patterns. I do not think either that it can be done by abstract inquiry into the philosophy of science, or by a review of the history of science. Some combination of these methods, applied to one of the active growing points of knowledge, might be more helpful than any single type of exposition in itself.

I think natural history is well adapted to this purpose. It represents a growing point of science, an area in which our ignorance is more impressive than our knowledge, yet an area in which our knowledge is constantly, hopefully increasing. It is an area of science of immediate concern to all of us, since we are animals, subject to its laws and observations, though at the same time we seek to apply those laws for our very human purposes. It is also an area of science that has not yet progressed to a level that can only be reached by special training in its symbols of thought. Yet its methods, its attitudes, its goals, are the same as those of any other science.

There are many definitions of science, corresponding to an equal number of ideas about what should be included under the term. Most broadly, science is systematized or organized knowledge. But it has come, particularly during the last century or so, to be used more and more in a narrower sense, for a special type of knowledge, arrived at by special sorts of activities.

The characteristic stamp of our Western civilization comes from science in this narrower sense. Ours is the "scientific" age, an adjective that is supposed to distinguish it from any other, though we can lay no claim to the exclusive development of systematized or organized knowledge. By science in this special sense, we mean particular kinds of study, like physics and chemistry and biology. And we are apt also to think that we mean the experimental method —though we readily include things like geology and astronomy among our sciences, despite the fact that they have few traces of experiments in their make-up.

For science in this present, narrow sense, I like the definition of James Bryant Conant. Conant says that "science emerges from the other progressive activities of man to the extent that new concepts arise from experiments and observations, and that the new concepts in turn lead to further experiments and observations."

Science is a progressive activity. The outstanding peculiarity of man is that he stumbled onto the possibility of progressive activities. Such progress, the accumulation of experience from generation to generation, depended first on the development of language, then of writing and finally of printing. These allowed the accumulation of tradition and of knowledge, of the whole aura of cultural inheritance that surrounds us. This has so conditioned our existence that it is almost impossible for us to stop and examine the nature of our culture. We accept it as we accept the air we breathe; we are as unconscious of our culture as a fish, presumably, is of water.

The equipment with which we face the world is of two sorts: our biological inheritance and our cultural inheritance. Within the time span of history, our biological inheritance has not changed. The physiology of reproduction, for instance, is the same as with our cave man ancestors—though the antics that precede copulation may be different now, and may vary greatly from culture to culture. Our digestive system remains the same, despite its varying degrees of maltreatment. The shape of our noses has not changed since Caesar's day. And while lately we have been growing taller in the United

States, this seems to be a result of better diet rather than of a shift in our biological potentialities.

Our cultural inheritance, when compared with our biological inheritance, seems to be subject to constant change. Though when we examine it carefully, we find that the rate of progress of its different parts is far from uniform. To realize this, we have only to compare our culture, our activities, with those of some other period in the line of historical antecedents. With Periclean Athens, for instance. Whether our activities in art, literature, philosophy, represent "progress" as compared with the Athenian is at least debatable. The amount of progress in politics or government might also be debated. We read the histories and wonder why the Greeks so stupidly clung to their petty city states, why their civilization was so persistently marred by internecine strife. But we seem still to be involved in the same sort of political jealousies, the same sort of armed strife, though we have changed the scale.

If we go back far enough, of course, we can detect all sorts of changes that have progressive aspects. If we compare ourselves with our cave man ancestors, we can see progress in art and philosophy as well as in mining and implements of war. But as we move along the time scale the progress in some things seems more regular than in others. The discovery of a technique makes possible the development of an art form, but the development proceeds in fits and starts. We might report progress in music, for instance, over Periclean Athens (though the evidence is pretty indirect), but this progress happened in a spurt after the discovery of harmony and the invention of a system of notation. The cumulative progress since Bach is at least not a matter of universal agreement. From this irregularity comes all of the discussion of the cycles of cultures and civilizations.

Along with these cycles of culture, we have the stream of progressive activities. Its first trickle is apparent in the discovery of fire, of ores, of agriculture, and its development can be measured by a constantly increasing power over the material universe. The stream has grown more rapidly in some periods than in others. But looking back over a large scale map of history, the stream seems surprisingly independent of the vicissitudes of cultures and civilizations. In Europe, even during the darkest part of the post-Roman Dark Ages, we find water mills developing, and the invention of things like stirrups and horse shoes. At that time, the main stream had left Europe in a curve that passed through the luxuriant periods of the Arabic and Indian civilizations, so that when it again turned towards Europe in the fourteenth and fifteenth centuries, it was much grander than when it left that continent with the collapse of the Graeco-Roman world.

Science can truly be said to emerge from this stream of progressive activities, but so imperceptibly that it is difficult to say exactly when, where or how. The history books trace its origins among the Egyptians, the Babylonians, the Greeks. It can be distinguished fairly clearly from philosophy at Alexandria, and it began to take definite shape during the bright period of Arabic culture. But as a strikingly distinct sort of activity, depending on the wide dissemination of individual discovery and thought, its development was hardly possible without the printing press. Its methods and attitudes are hard to define until we come to the first of the great modern scientists, Galileo, who lived in the sixteenth century.

Experiments, observations and conceptual schemes, if we accept Conant's definition, then form the basic components of science, of this progressive activity that first clearly emerges with Galileo. They must, in any particular science, act together as a self-maintaining system, an endless series in which new concepts are constantly formed or old ones modified through the factual accumulations resulting from the observations and experiments. The problem, in understanding any particular science, is to see how this modifying process works: how, in the present case, it applies to natural history.

One dictionary that I consulted remarks that "natural history" now commonly means the study of animals and plants "in a popular and superficial way," meaning popular and superficial to be equally damning adjectives. This is related to the current tendency in the biological sciences to label every subdivision of science with a name derived from the Greek. "Ecology" is erudite and profound; while "natural history" is popular and superficial. Though, as far as I can see, both labels apply to just about the same package of goods.

Natural history is not equivalent to biology. Biology is the study of life. Natural history is the study of animals and plants—of organisms. Biology thus includes natural history, and much else besides.

The world of organisms, of animals and plants, is built up of individuals. I like to think, then, of natural history as the study of life at the level of the individual—of what plants and animals do, how they react to each other and their environment, how they are organized into larger groupings like populations and communities. Other biological sciences take up the study at other levels of organization: dissecting the individual into organs and tissues and seeing how these work together, as in physiology; reaching down still further to the level of cells, as in cytology; and reaching the final biological level with the study of living molecules and their interactions, as in biochemistry.

No one of these levels can be considered as more important than any other. The phenomena at each level are different, and we must try to get an understanding of each. A cell is something more than an aggregate of molecules; an individual more than an aggregate of organs. A population or community, for that matter, is something more than an accumulation of individuals.

Studies at all of these levels can be fun; and can be important from the human point of view of trying to manipulate the universe for our own comfort, and of trying to understand it for our peace of mind. I happen to enjoy most the study at the level of the individual; and this level is also, I suspect, most easily understood. I am an individual; so are you; so is my dog and the oak tree on my lawn. What makes us act the way we do? How do we get along with each other—the oak tree, the dog and the man? How did we come to get this way?

The physiologists, cytologists and biochemists are very largely involved with the problems of explaining living systems in chemical and physical terms. This is surely one of the major objectives of biology, but it need concern us little in our pursuit of natural history. The explanation of the living process is one thing; the explanation of the diversity of living things is another; and the latter may well be taken as the major objective of natural history itself.

The diversity of life is extraordinary. There are said to be a million or so different kinds of living animals, and hundreds of thousands of kinds of plants. But we don't need to think of the world at large. It is amazing enough to stop and look at a forest or at a meadow—at the grass and trees and caterpillars and hawks and deer. How did all of these different kinds of things come about; what forces governed their evolution; what forces maintain their numbers and determine their survival or extinction; what are their relations to each other and to the physical environment in which they live? These are the problems of natural history, problems that concern us ourselves as animals and that concern us even more as originators of this thing we call civilization—which is, after all, merely a rather special sort of an animal community.

The explanation of these things furnishes us with an objective; one that at times seems hopelessly remote. But whether we ever get to the objective or not, we keep finding interesting things along the way so that the going, however difficult, never seems tedious.

CHAPTER II

The Naming of Organisms

CHARLES ELTON has remarked that there is little use in making observations on an animal unless you know its name. The first step in a survey of natural history, then, should be the acquisition of some familiarity with the system of names and the system of classification, with the word equipment used by naturalists.

Many animals and plants have vernacular names that everyone learns in childhood, or that form parts of special vocabularies such as those of farmers, woodsmen or hunters. It is surprising, though, how quickly we exhaust this supply of names. It works well enough for large mammals such as bobcats, deer, foxes, raccoons. But if we start to make observations on field mice, we soon find that there are no common names for all of the kinds that we find; in the ordinary course of events, these different kinds of field mice simply don't come to our attention.

If we start to study insects, anywhere in the world, we find almost at once that our ordinary vocabulary is of no use: at most it gives names for a few types. With plants the situation is similar, though the list of common names for conspicuous or useful plants is probably longer than the animal list. The vast majority of animals pass unnoticed in our daily lives unless they happen to annoy us. Then we swat them or step on them, and no name is needed.

Everyone nowadays knows something about the technical names that biologists use. These names come in handy if we want to add a touch of humor, as by some elephantine reference to Homo sapiens. Gardeners are apt to be very fluent with the names for different species of Iris or Gladiolus or whatever happens to be their specialty. Physicians all have a speaking acquaintance by Latin name with a great many bacteria, protozoa and parasitic worms. Quite a few technical names, especially for garden plants and zoo animals, have passed over into everyday usage.

Yet most technical names still have a queer, foreign, pedantic sound, and if we see very many of them in a stretch of print we get frightened off and hastily look for something else to read. This attitude is sustained by the typographical habit of putting such names in italics, to indicate their foreign origin. But the technical names are not foreign to any language. They are common to all languages. They will not be italicized in this book: it is a detail, but perhaps it will help readability.

Vernacular names have a limited usefulness for biological purposes. In the first place, they are available for only a few kinds of organisms. Second, they are apt to be very local. Some names have gained wide currency in English because of literary practices, so that we do not realize how regional a vocabulary can be. Even in English an animal like the puma may have quite a list of aliases—catamount, cougar, lion, panther. In some countries all of the names for common animals and plants may change from valley to valley. Third, at very best the names are limited to one language, and science must necessarily strive after internationalism in its vocabulary as well as its ideas. There are millions of different kinds of organisms, and the invention of parallel millions of names in each language is unthinkable. Fourth (I think this list is about long enough) it is hard to give a vernacular name a precise meaning. Words like oak, or pine, or rabbit, may cover a variety of things, and we can't always straighten this out with adjectives like white oak or cottontailed rabbit. We often (perhaps better, usually) discover inconspicuous but significant differences that would make it necessary to name the Florida cottontailed rabbit, or Wilberforce's eastern white oak, and in the end we would find ourselves completely and hopelessly tangled up in our vernacular vocabulary, which started out looking so simple.

Biological nomenclature forms a beautiful system. I doubt whether it is fully appreciated even by the average biologist, who is apt to be irritated by the trivial inconsistencies that turn up through daily contact. I have heard biologists who envied the chemists their names. Sodium chloride and carbon dioxide may be all right, but my envy of the chemical names stops when I look at the fine print in the medicine cabinet. Sulphathiazol, for instance, is the vulgar name for 2-(para-aminobenzolsulphonamido) thiazol. We haven't got anything as bad as that in biology; and the chemists really have a hard time when they start applying their system to something moderately complicated like the proteins. On the other hand, the chemical names do indicate the structure and hence affinities of the chemical, while biological names are arbitrary symbols.

THE LINNAEAN SYSTEM

Biological nomenclature is the invention of Carolus Linnaeus. I have just looked him up in the encyclopedia, and find that he gets slightly less than a column of biography. It is curious to find that the author of one of the great achievements of the human mind gets so little attention. On looking further, I find that Isaac Newton, surely one of the greatest of men, gets only half the space of David Lloyd George. It isn't specific neglect of Linnaeus, it is lack of general interest in the biographies and characters of the men responsible for modern science.

Linnaeus, a Swede, was born in 1707 and died in 1778. He was the eldest son of a peasant, Nils Ingemarsson, who became a pastor and who, with this rise in the world, adopted Linnaeus as a family name from a huge linden tree that grew near his home. Carl apparently even in childhood showed a great interest in plants, which in those days meant mostly medicinal herbs, and he resolved to study medicine. He was very poor, and probably would have been unable to survive at all at the university of Upsala except for the help that he early received from Celsius, the dean. Linnaeus must have been a very engaging fellow, since all through his life he received the admiration, friendship and sympathy of his fellow scientists. Perhaps he had that quality of infectious enthusiasm. Certainly he must have been an enthusiast, since otherwise he could never have carried through the tremendous volume of work for which he is responsible.

He received various grants and honors at Upsala, but in order to obtain the degree of doctor of medicine, he had to leave Sweden. So, with funds advanced by his future father-in-law, he went to Holland where he stayed for several years. With the assistance of patrons in Amsterdam and Leyden, he published the first edition of his great book, the *Systema naturae.* This was in 1735, when he would have been twenty-eight years old. He published other books in fairly rapid succession. He returned to Sweden in 1738 and continued to live there the rest of his life, refusing honors and appointments abroad. He was, however, held in high esteem in his own country, where he had the professorship first of medicine and then of botany at Upsala, and where he was given noble rank. He had large numbers of students whom he sent on collecting expeditions to various parts of the world.

The work of Linnaeus that interests us most is the "System of Nature." Twelve editions were published during the author's lifetime, each showing progressive changes and additions. The tenth edition was published in 1758, making this one of the basic dates of biology, like 1066 in English history, or 1776 in American history, for this tenth edition is the starting point, the cornerstone, of zoological nomenclature (plant names date from the *Species Plantarum* of 1753). This edition contains the first full expression of the binomial system, in which every kind of animal and plant is given two names, the first generic, showing the group to which the organism belongs, and the second specific, giving a name to that particular species.

Linnaeus divided the Empire of Nature into the Kingdoms of Animals, Plants and Minerals. The Kingdom of Animals he divided into six classes: the mammals, the birds, the amphibia, the fish, the insects and the worms. The classes were in turn divided into orders, the orders into genera, and the genera into species. Linnaeus went through the material of all of the museums available to him, studied all of the books on animals and plants that he could get hold of, and gave a name in his system to every kind of organism thus known to exist, with a brief description of its characteristics, and with page references to the places where it was mentioned in the various natural history books. Even aside from the invention of the system of naming, this represents a remarkable achievement of industry, of critical evaluation, by a brilliantly systematic mind.

Linnaeus was engaged in making a catalogue of fixed species, each separately created. He was, however, also trying to develop a natural system, by which organisms closely related to each other would be classified in the same groups. He was primarily a botanist, and a great deal of his life was spent in trying to work out a natural classification of plants, a goal to which he came closer in each of his works. His classifications have, of course, undergone many changes over the years: but after all, his great contribution to science was the invention of a system that permitted orderly change with the constant increase in knowledge.

There is no point in including more material here on the history of biological nomenclature, beyond mentioning one event—the publication, in 1859, of Darwin's book, the *Origin of Species*—because the meaning of biological names was quite different before and after 1859. During the hundred years between the tenth edition of Linnaeus' "System of Nature" and Darwin's *Origin of Species,* the problem was to make a catalogue of the fixed, unchanging kinds of animals and plants that were created for the populating of the earth. After 1859, the problem became one of recognizing various stages in the family tree of animals and plants, of separating out more or less artificial steps in material that was subject to constant change, to evolution.

I have, of course, oversimplified, since many naturalists were aware of various possible sorts of evolution before 1859, and since many persisted in regarding species as something fixed by God for quite a few years after 1859. Even so, the revolution caused by that one publishing event was probably more rapid and more drastic than that caused by the publication of any other single book in the history of man.

THE CONCEPT OF SPECIES

The basis of biological names is the *species.* A species is a kind of animal or plant. Nowadays we like to think of a species as a population which includes all of the individuals that, in the natural course of events, could mate or are likely to mate with one another. All of the bizarre varieties of domestic dog still constitute one species, because they all belong to a general inter-mating population, as everyone knows who has tried for a few days to protect a bitch in heat. Red foxes and timber wolves do not look more different than many varieties of domestic dog; but red foxes do not normally mate with timber wolves, hence they are different species.

I shouldn't have got started with dogs, because they are a rather special case. No one is sure what wild dog, or which wild dogs, formed the ancestry of our household pets, and domestic dogs can be crossed with various kinds of wolves in different parts of the world. The important point is that they do not habitually cross, so that in actual fact we have a number of reproductively isolated populations, which we can safely call species.

As long as we stick to one place, we can be fairly sure about species. If, around our town, we have three kinds of mice and two kinds of squirrels; if we observe that there aren't any or very many intermediates among these mice and squirrels; and that they have somewhat different habits as well as a different appearance, we can be pretty sure that we have three species of mice and two of squirrels. If the males of one kind of squirrel took to chasing after the females of the other kind with any frequency or with any success, the place would soon be overrun with a lot of intermediates so that our two original kinds would no longer be easy to distinguish—in fact, they wouldn't be two species.

The naturalist is most apt to get in difficulties when he has a squirrel from Utah and another one from Massachusetts. Maybe one has more gray in his tail than the other, a few more whiskers, and slightly bigger ears. Now, are these two species, or are they varieties of the same species? You can't always get a live male from Utah and a live female from Massachusetts to try it out; and anyway, behavior in captivity is rather uncertain. Sometimes animals that wouldn't touch each other with a ten foot pole in the wild will carry on scandalously if they are put in a cage together. Other animals that have no inhibitions out in the woods take monastic vows as soon as they are put in a cage. So it isn't easy to use the experimental method.

What the naturalist actually does is to try to get a lot of specimens of squirrels from country in between Utah and Massachusetts. If the change seems to be gradual, the tail getting grayer, the whiskers more luxuriant, the ears bigger as Massachusetts is approached, he calls his two squirrels subspecies (geographical varieties, or races if you will). If the change is abrupt somewhere about Missouri, he assumes that they are two species: that somewhere along the line they have had a chance to mix and have turned up their respective noses. If the naturalist can't get enough specimens to decide about this, he guesses, and some of his colleagues agree with him, and others don't.

Thus the question of what is a species is really pretty much a matter of judgment by experts, and attempts at definition or at the establishment of objective criteria have not met with much success. In the vast majority of cases there is no problem, and the experts are pretty much in agreement. The cases where the experts are not in agreement are perhaps the most interesting, since the study of such cases may provide clues to the problem of the origin of species. Sometimes there is a maze of forms that is so complex that no two people agree on how to sort them out into species. The North American blackberries (Rubus) are a fine example. Hardly any two wild blackberry plants are alike, and some people think there are hundreds of species, others that there are only a few. Perhaps in this case man has caused the confusion by his drastic alteration of the landscape. When North America was covered by forest, blackberries may have been rather rare, limited to small open areas. With the clearing of forest, their opportunities for spread increased enormously, so that a few originally separate types came in contact, hybridized, and produced an endless range of intermediate forms.

Man himself provides an interesting taxonomic problem (sorting things out into their proper species is part of the science called taxonomy). We can see, with man, the geographical pattern of various physical types: Negroes in Africa, Caucasians in Europe, Mongolians in Asia and so forth. We can also see that these various types are but variants of one species, since complete mixture occurs wherever two types are long in contact. There has, in fact, been so much contact through migrations and other types of population movements, that the anthropologists are unable to come to much agreement in attempts to classify man into subspecies (or races). Their problem is much like that of the botanists faced with the North American blackberries. Some anthropologists make dozens of races, while others can see only four or five. Some advocate abandoning the concept of race altogether.

Despite these confusing cases, the idea of species corresponds to something that has real existence. At a given time and a given place—such as around your town during this century—animals and plants are divided into discontinuous populations, isolated from each other mostly by their habits of reproduction. An important task of the naturalist is to catalogue all of these populations, giving them each a specific name, so that he can study them and communicate the results of his study to others.

THE GROUPING OF SPECIES

Names for species can hardly be handled unless they are classified into groups. The more natural the groups are, the easier will be the problem of handling the names (and the corresponding organisms). Just as individual plants and animals fall into discontinuous populations, species, so the populations fall into groups the members of which are more or less similar to each other in various ways. The problem is to arrange a hierarchy of groupings that will be both convenient and natural.

Linnaeus, as I said before, grouped his species into genera, those into orders, and the orders into classes. We have added several categories to this hierarchy. Among animals, we have now the species, the genus, the family, the order, the class and the phylum, sometimes with the tribe placed between the genus and family. Each of these may be further split with "sub" or "super" categories, so that we can have subfamily, family, superfamily, suborder, order and so forth.

THE GENUS

A group of similar species forms a genus. There is no rule about how a genus is formed and the genus doesn't correspond to any definite thing in nature—it and all of the higher categories are primarily conveniences, necessary filing systems for our accumulating information. The word genus keeps its Latin plural in English, so we write about various genera. Generic names are always capitalized, and the generic and specific names go together to make up the name formula for the organism. Take cats, for instance. The cats in general form the genus Felis. The house cat is Felis domestica. Like all of our domestic animals, its origin is a little uncertain, but it is supposed to be descended from the Egyptian wild cat (Felis ocreata) and the European wild cat (Felis catus), various breeds perhaps having different proportions of each ancestry.

A little while ago I wrote that the idea of species corresponded to something real in nature—to discontinuous populations—and here I pick another example where the thing is obviously artificial. Keeping Felis domestica, F. catus and F. ocreata separate is pretty much merely a convenience if the first is a mixture of the other two. But biologists are not always consistent, and the separation of these three kinds of cat really is a convenience. The domestic animals, in any case, present rather special nomenclatorial problems. The lion, Felis leo, and the tiger, Felis tigris, are better examples of species. The American jaguar (Felis onca) is another. The North American bobcat is generally placed in another genus, and called Lynx rufus; the cheetah (Cynaelurus jubatus) is another example of a cat that is usually considered not to belong to the genus Felis. All of these cats, though, are put together in the family Felidae, which takes us another step up in the hierarchy.

The genus and the species form the basic name formula of an animal or plant and these two names are all that is needed to look up what has been written about the organism (if the library is big enough). This is accomplished by two rules: that a particular name can be used for a species only once in a particular genus; and that a particular name can only be used once in the animal or plant kingdom for a genus. A word used for a genus of plants can be used again as a generic name in animals, and there are a few cases; but since it is usually fairly obvious whether a plant or an animal is being discussed this causes a minimum of confusion. It is customary to capitalize the generic name but not the specific name, though some botanists capitalize specific names if they are proper nouns.

There are tens of thousands of generic names in use for animals, so that keeping track of these names, to be sure that they are never used twice, is a large job. Various lists have been compiled of all of the names that have been proposed and the Zoological Society of London publishes every year a list of all new names of animals. The botanists have a similar system.

It is surprising how difficult it is to think up a word for a new genus that no one has used before. I may think I have a combination of letters that could not possibly have been thought of, but when I start to check back, I find that John Smith used that word for a genus of fish in 1897, or Everett Wilberforce for an Australian beetle in 1912. That is one reason why so many scientific names turn out to be jawbreakers: the theory is that the bigger the jawbreaker, the less likelihood that someone else will have thought of it. The committee on rules for zoological names has tried to put a stop to this by setting a limit on length, but their limit is still rather long. Another reason for jawbreakers, of course, is that some scientists like them, probably because they have a learned sound.

A genus may have any number of species. Sometimes a genus has only one species, if that particular kind of animal or plant has no close relatives. Thus man is put in a genus by himself among living animals, though various fossils have been described as other species of Homo. Some genera have hundreds of species: groups of animals or plants where there are many slightly different kinds, some perhaps quite different from others, but with such gradual connecting links that it is most sensible to include them all together under one generic name.

It is important to remember that all of these categories above the level of species have an artificial basis in convenience. I have worked quite a deal with two families of insects—one the mosquitoes and the other hawkmoths. In the mosquitoes there are on an average about forty species to a genus, in the hawkmoths five species. About two thousand different kinds of mosquitoes are known, and almost as many hawkmoths. I don't think the evolution of the two groups has been very different. It just happens that the people who work with mosquitoes, following the lead of one or two specialists, have got in the habit of using large genera, while the people who work with hawkmoths like small genera. Personally, I like the larger genera, because it makes the names easier to recognize, but there are many arguments for the small genera. Some people split the difference and use big genera with many subgenera. They then write the subgeneric name in parenthesis, like Anopheles (Myzomyia) gambiae, which happens to be the mosquito primarily responsible for malaria in many parts of Africa.

FAMILIES AND HIGHER GROUPS

Families, groups of genera, are also sometimes very big and sometimes very small. Among animals, family names are always made by adding the letters "idae" to some genus name. Thus the dogs are Canidae; the cats, Felidae. In plants family names are usually formed by adding "ceae" to a genus: Rosaceae for the rose family and Liliaceae for the lily family. Scientists are apt to differ considerably in defining the extent of families, and again it seems to me that the final decision should rest on convenience as well as on the expression of natural relationships. Families are rarely covered by words in the common language. It is necessary to refer to the Canidae, for instance, as the "dogs and their relatives", the Procyonidae as the "raccoons and their relatives", the Liliaceae as the "lilies and their relatives" (including the onions).

Groups of families make orders, and orders are more apt to be covered by common words than are families. Thus the beetles and flies represent orders of insects, the primates, carnivores and marsupials orders of mammals. Plant orders, curiously, are rarely recognized as discrete groups by the non-specialist. Orders again are grouped into classes, and the distinction of the classes is often obvious enough so that the groups are recognized in the common vocabulary. Thus the classes of chordates (vertebrates) include the mammals, birds, reptiles, amphibia and fish. Insects form a class of arthropods. The major subdivisions of the animal kingdom, such as the chordates (vertebrates), the arthropods, the molluscs and the protozoa, are called phyla. The term "phylum" is less often used by the botanists.

The classification of organisms is built up from individuals. The specimens studied by the naturalist are grouped into species— which is already an abstract "concept", though one that corresponds to a fairly concrete phenomenon insofar as species are interbreeding populations. Genera are defined on the basis of a study of the characteristics of many different species, families on genera, orders on families, and so forth.

In using this classification, however, the naturalist works the other way. Given a particular organism to identify, the first question is, what phylum does it belong to? Is it a seed plant, a moss, a vertebrate, a protozoan, or what? The next question is, what class does it belong to? Which, among the vertebrates, means deciding whether it is a fish, amphibian, reptile, bird or mammal. Within the class, the order, then the family, then the genus, and finally the species must be determined. When these last two have been found, the name formula of the organism is at hand.

The determination of family, genus and species generally requires a rather detailed knowledge of the anatomy of the particular phylum of organisms involved. Since the basic anatomical structure of the various phyla differs greatly, this is a specialized knowledge. No biologist would attempt to identify all of the organisms he comes across. He usually acquires an intimate knowledge of one or a few groups, and when he wants to know the name of some other kind of organism, sends it to some friend who specializes in that group. The naming of organisms is thus a co-operative enterprise, and forms the chief function of the great museums where specialists for the various phyla and classes are gathered together, with large reference collections of preserved material.

These specialists are the archivists who maintain the catalogue of organisms, thus providing the reference frame for the accumulation and relation of the observations of all of the other varieties of naturalists. Our next step in exploring natural history, then, should be to examine the main divisions of this reference system for the catalogue of nature, which will serve also as a rough sketch of the diversity of these basic kinds of organisms.

CHAPTER III

The Catalogue of Nature

IT is usually estimated that about a million species of organisms have been described and given names. No one can make more than a rough guess as to how much progress this represents toward the goal of getting all of the kinds of organisms named. In a few groups, like the birds, almost all of the very distinct kinds have surely been found and catalogued. In other groups, like some of the smaller and less conspicuous insects, only a small percentage of existing kinds have been given names. Charles Brues has estimated that there may be ten million species of insects alive in the world today, and that only a half a million or so of these have as yet been catalogued. Whatever the exact figures, it is clear that the job of naming is enormous. Biology has got swamped by the problem, and the end is nowhere in sight.

The catalogue of nature interests us, as naturalists, chiefly because it is a necessary tool. Without some such filing system, the accumulation and indexing of observations and experiments would be impossible. Imagine trying to keep notes that apply to one kind of thing out of a million or so, without any generally accepted filing system.

The last chapter was devoted to the theory of the naming system. Before going on to the general problems of natural history, to the description of the behavior and interrelations of organisms, it may be useful to outline the system of classification. The present chapter, then, is meant to give a bold sketch of the major types of animals and plants that have been found living in the world today. I shall try, in the next chapter, to give this classification historical perspective by a review of what we know of the geological history of these major types. This should provide useful background for the rest of the book, which will deal with particular sorts of processes, behavior and relationships shown by organisms.

PLANTS VERSUS ANIMALS

In forming a classification, we have to list our groups in a lineal order. This, inevitably, is arbitrary. We may compare the evolutionary history of organic types to a branching tree, the various sorts living today being the end result of this historical process of divergence, the final twigs on branches that have spread in different directions during the slow course of growth in geological time. To make a list of groups, we have to cut these branches and lay them out in a straight order, A, B, C, D and so forth, which may give a very misleading impression.

The basic fork in our tree, for instance, involves the separation of plants and animals, a divergence that started somewhere very far back in geological time. It is easiest to deal, in the list, first with all of the plant groups, then with all of the animal groups. But this doesn't mean that plants are "lower" than animals, or ancestral to animals. They are two forks of a tree that have got to be dealt with somehow in a list on a sheet of paper. The order on the list is a pure convenience, except that we try to group the branches in a way that reflects our idea of how the main trunks probably diverged on this hypothetical tree back in geological time.

The difference between plants and animals is taken as a matter of course with complex organisms—anyone can tell a horse from an oak tree. But with relatively simple organisms, the differences are not so obvious, and in some cases it is really impossible to decide whether a given organism should be called a plant or an animal.

We think of plants as fixed, growing in one position, and of animals as capable of various kinds of movement, of locomotion. There are many exceptions, however. Quite a variety of marine animals have fixed positions—corals, sponges, sea anemones, for instance. Moving plants are less common, but the slime molds creep, and some microscopic algae are very active.

The basic difference between plants and animals is one of food economy and habits. Plants are the key industry organisms on which the development of other forms of life depends. They can start with carbon dioxide from the air and through the magic of green chlorophyl build up starch and from this the complex molecules of carbohydrates and proteins, which animals must get by eating plants or by eating other animals that in their turn have eaten plants. As for habits, animals eat their food, while plants absorb their food in water solutions through cell surfaces. Even microscopic animals live by engulfing pieces of food, while microscopic plants lie in the sun and absorb the surrounding liquids. The exceptions to this are mostly parasitic animals that have taken to absorbing predigested food in the intestines or blood stream of other animals. It is largely this difference in food habit that leads biologists to include the great group of fungi, including bacteria, among plants, since these organisms do not possess chlorophyl, though they are capable of some astonishing chemical tricks.

VIRUSES

But if we start the filing system with structurally simple organisms, the first group is one that I have never seen even the most enthusiastic zoologist list among the animals, or the most grasping botanist among the plants. This group includes the viruses, things so small and so simple that no one understands them. So small that they cannot be seen through a microscope with visible light because they would fall between the light waves; so simple that it is possible to have long arguments about whether they are really alive at all or not.

W. M. Stanley was given a Nobel prize because he was able to turn one virus—the cause of a plant disease called tobacco mosaic—into crystals, and to prove that his crystals were really the virus, the cause of the disease. Now many complex chemicals may be crystallized, but it is hard to imagine turning a living organism into a crystal, or a lot of crystals, and then turning it back again. Yet this tobacco mosaic undoubtedly grows, multiplies, reproduces itself, acts in most ways like a living thing.

All known viruses are parasites. The parasitic habit will come in for a deal of attention later in this book, but let us assume now that everyone knows what a parasite is—an organism that gets inside of another organism, or hangs on the outside, and makes it sick, or at least makes it want to scratch. The trouble is, we would have no way of catching a virus if it were not a parasite, because we can recognize the virus only by the symptoms that it causes. Virus particles can be photographed with an electron microscope, but it would have been impossible to get the stuff to make the picture if the virus hadn't made some animal or plant sick to start with. There may be all kinds of harmless viruses lying around in the mud, but we have absolutely no way of finding them, of distinguishing them from all of the dead contents of the mud.

It is possible that all viruses are parasites—that they can be so small because they depend on their host for everything and manage themselves only to be the very essential life stuff, the primordial chemical nucleoprotein, which keeps on reproducing, multiplying, as long as it has host cells to do all of the work. All parasites become, through the eons of geological time, increasingly lazy as they discover how easy it is to make their hosts do all of the work. Perhaps viruses are the last end of such an easy life. Perhaps in ages past they were fine, upstanding, complex bacteria, filled with all kinds of chemicals, and able to look after themselves quite nicely, even making their own way in the world. Perhaps they got smaller and smaller and simpler and simpler as they came more and more to depend on generous hosts until finally, in extreme cases, nothing was left but that last chemical, that last essential component that can only carry itself on, and that Stanley so brilliantly turned into tiny, needle-like crystals of nucleoprotein.

BACTERIA

Which brings us to the bacteria. The botanists always list the bacteria very boldly among the plants, calling them Class Schizomycetes of the fungi. The botanists, however, pay little attention to bacteria beyond listing them in the books; and the bacteriologists, who actually study bacteria, usually shrug off the problem of whether to call them plants or animals with a few paragraphs pointing out ways in which they might be classed as either, or neither. They are very special organisms that must be studied by very special methods.

Bacteria are defined as organisms built of a single cell, without apparent nucleus. Bacteria are of all sizes and shapes—though a microscope is still needed to see the biggest one. They may include various kinds of things that are not really related, and one group (the corkscrew spirochaetes of syphilis) gets pushed back and forth between the animal and plant kingdoms, each author having his own ideas on the subject. The bacteria as a whole have got a bad name because of a few kinds that cause diseases, but the vast majority are busy doing their microscopic but tremendously important part to maintain the economy of nature. Bacteria are the basic organisms in the process of rotting the corpses of dead animals and plants into simple materials that can be used again. Many kinds of bacteria have taken up partnership with big organisms, aiding in the digestion of things like cellulose in return for a safe and protected refuge in the intestine or in other parts of the big partner.

Perhaps the most important bacteria of all are the ones that keep the supply of nitrogen constantly going for the rest of the organic world. Some of these live free lives in the soil, others have taken up partnership with bean plants (legumes); each kind has specialized for a particular operation in the nitrogen chain, and the whole thing makes a fascinating story that has been fairly well publicized, though perhaps not well enough. There are other bacteria that can do tricks with sulphur and phosphorus; bacteria that get along very nicely without air; bacteria that are not killed even if they are boiled. There are bacteria everywhere; they are by far the most ubiquitous and numerous of organisms.

FUNGI

Next come the fungi—mushrooms and a lot of poor relations. The poor relations have been coming up in the world, though, since one of the molds, Penicillium, got into the newspapers and the medicine cabinets. Bacteria are usually considered as a division of the fungi, which would then include all plants that do not have chlorophyl; other divisions of the fungi are the yeasts, molds, and various types that do not have definite group names in the common language. I have been looking up several classifications while writing this chapter, and I can't find any two classifications of the "lower plants" that agree in details of arrangement.

The phenomenon of sexual reproduction appears with the fungi. Knowledge of sex among the bacteria is scanty and indirect, but with fungi sex is well studied and has all kinds of complications, since the same plant may look quite different when it is in a sexual phase from the way it looks when going about its ordinary asexual business of rotting shoes. Yeasts are the simplest fungi, but even they present sexual problems—in fact the book open before me lists four different ways in which yeasts can multiply.

Yeasts are important in the chemical economy of nature. I think we all know something about Pasteur's discovery of the role of yeasts in fermentations. For the rest, suffice it to say that there are a great variety of fungi, from simple yeasts to mushrooms and puff-balls, including a great many inconspicuous types, sometimes para-sitic, sometimes playing important parts in natural economy, espe-cially in the maintenance of soil conditions favorable for other organisms.

ALGAE

The algae are the simplest of the green plants, the plants with chlorophyl that can make starch out of air and water and sunlight. Almost all algae live in water, and every housewife knows the kinds that form a green film over the sides of a glass vase that has been left too long with water. Aquarium enthusiasts have rubber gadgets for scraping the algae off the sides of their aquaria so that they can see the fish. Even the clearest of natural waters have a consid-erable population of algae, all microscopically busy making starch, waiting to be eaten by microscopic animals that will in turn be eaten by somewhat larger animals, and so on up the scale of size. They make up, then, the pastures of our ponds and seas.

The algae, like the fungi, are of all sizes and shapes. They are normally divided into a variety of classes depending on their pre-dominant color—the green, the yellow-green, the brown, the red and the blue-green algae. A great many are single-celled organisms, and some of them, because of their active movement and close similarity to the animal protozoa, are claimed by both the botanists and the zoologists. Some microscopic algae (diatoms and desmids) have beautiful symmetrical forms, like the patterns of snow crystals, and are favorite objects of the microscopists. In others the cells group together in various ways, first forming simple chains of cells, then forming more complex communities with different cells taking on different functions, until finally many algae form definite, obvious plants, the seaweeds, including the largest plant of them all, the giant kelp of the Pacific ocean. These algae never, however, achieve the special structures of the seed plants, roots, stems, leaves, flowers, though certain parts of the plant may look like a stem, a root, or a leaf.

LICHENS

I suppose lichens should be the next group of plants on our list. We all know them, making splotches of color on rocks, or forming tiny Japanese gardens on old logs. These lichens are always listed as a special class of plants, though in reality they are not a single class of plants at all, but a queer natural partnership in which the thing that we see is a mixture of two completely different kinds of organisms, of a fungus and an alga. Each kind of lichen is made up of a particular kind of fungus and a particular kind of alga. The fungus provides support, salts and water, while the alga carries on its business of manufacturing starch. This partnership results in a curious naming situation: each kind of lichen has a name as a lichen, yet it is composed of a particular kind of fungus with a name in the fungus system, and of alga with a name in the algal system. In some cases the algae and fungi are of kinds that can also live separately, but mostly the fungus, at least, cannot live without its alga. The partnership involves quite different types of algae and fungi in different types of lichen, and must thus be a very ancient agreement. Perhaps the different types of fungi discovered quite independently the advantages of having some kind of alga always close at hand.

MOSSES AND LIVERWORTS

The mosses, with their cousins the liverworts, form the next great group of plants, called Bryophyta by the botanists. The algae are essentially aquatic organisms, and the mosses are the simplest green plants that have solved the problem of how to grow on land.

Even with mosses, the problem has been only imperfectly solved, since they are pretty much limited to damp places. In this group the tissues of the plant have become differentiated into root and stem, and the liverworts have leaves, but these parts do not really correspond to the highly specialized root, stem and leaves of the higher plants. I haven't said anything about reproduction in fungi and algae, beyond mentioning that these plants get involved with sex, because the reproductive mechanisms are very diverse and complex. With mosses, too, reproduction is a complicated business of spore formation and alternation of sexual and asexual generations.

FERNS

With the next major group, the ferns, we meet plants in the conventional sense of the word. The ferns have solved the main problems of growing on land, but they still have a reproductive cycle reminiscent of the algae and fungi. They do not have flowers or seed, but produce millions of spores, of special cells, usually in little brown packets on the under side of the leaves. These spores are produced in tremendous quantities, so that a single plant may release fifty million or so in a season. The spores grow into tiny, scale-like plants, called prothalli, which produce sexual cells, the sperm cells swimming through water from dew or rain to reach the female cells, a lingering trace of the aquatic life of the simpler plants. The fern that we know grows from this fertilized female cell, so that there is an alternation of generations of sexual and asexual plant forms.

Ferns were dominant in the landscapes of the geological past, and there are many fossils in different kinds of rocks and in coal beds. Eight or nine thousand species have been found growing in the world today, some of them, in the tropics, making fair-sized trees.

This business of numbers is interesting. Some 14,000 kinds

of algae have been described; it is estimated that 100,000 species of fungi have been described; of bacteria, the number described is of no importance since it must be only a minute fraction of the number of living kinds; of mosses and liverworts, some 17,000 species are known. But the dominant plants of the world today, both in number of kinds and in bulk, are the next group, the seed plants, with something like 150,000 species.

Before getting involved with the seed plants, I should mention that the ferns belong to a group (the Pteridophyta) with two other kinds of plants that still persist inconspicuously in the world today after a glorious geological past—the horsetails and the club mosses.

SEED PLANTS

There is probably no use in writing very much about the seed plants here. We all know the more obvious things about them, remembered from high school botany or picked up at meetings of the Garden Club. My chief object has been to put the seed plants in perspective, by listing them with the other major plant groups.

The seed plants, which may be called phanerogams or spermatophyta, are divided into three classes, all with jawbreaker names, the gymnosperms, the monocotyledons and the dicotyledons. The gymnosperms do not have flowers in the usual sense, the seeds being borne on scale-like leaves—in a pine cone, for instance. These plants were abundant in the geological past, and the coniferous forests still cover a respectable part of the earth's surface.

From a grain of corn, a single leaf-like shoot comes out of the ground; from a bean, a nicely balanced pair of seed leaves first spread out. This difference is the key to the division between the monocotyledons and the dicotyledons. Both are divided into a series of orders and numerous families, the monocots leading up through grasses, palms, lilies and bananas to reach their most complex development in the orchids. The dicots include all of our ordinary herbs, shrubs and trees, and are considered to reach their highest or most complex development in the composites, plants of the daisy family.

I started out this catalogue with the tree analogy, whereby diverging evolutionary history can be compared with a branching process. In that figure, we have got clear out at the tip of the plant fork with the orchids and the daisies, and we now have the problem of climbing back down to the main trunk, to the world of single-celled organisms, to start out again on the animal fork with the protozoa.

The protozoa are one-celled animals. Among plants the bacteria and many types of algae are single-celled, and of course all organisms pass through a stage—the fertilized egg—in which the individual consists of only one cell. Cellular organization has received a great deal of attention in biology. Robert Hooke, late in the seventeenth century, first noticed the microscopic division of various plant structures into compartments which he called "cells." Observations on cells were accumulated by various workers, including Caspar Wolff, who first noticed the similarity in cellular development between plants and animals. Two Germans, Matthias Schleiden and Theodor Schwann, living in the first part of the nineteenth century, developed these observations into one of the great generalizations of biology, that all living things are made of cells. Cells since that time have been one of the prime subjects of biological study, involving two special sciences, cytology and histology, and every elementary textbook contains a great deal of material on cell structures and functions.

In complex animals, the cells are specialized for particular functions, but in protozoa all of the functions of the animal—digestion, excretion, locomotion, respiration—must be carried out within the cell, and various special functions may be performed by special parts of the cell which act like organs of complex animals, and so are called "organelles" or "organoids." The fact that protozoa are single-celled, then, does not necessarily mean that they are simple; and it has often been suggested that they would better be called noncellular animals rather than unicellular animals.

The protozoa, from any point of view, are fascinating creatures, and their study (protozoology) has become an extensive and specialized science. Most protozoa are quite active, some creeping in a manner that has become familiar to us all as "amoeboid movement", some moving rapidly with the beating action of numerous tiny hairs (cilia), or by the lashing of a single long tail (flagellum). Many have become parasitic and several important human diseases (malaria, African sleeping sickness, amoebic dysentery, for instance) are caused by animals of this group.

The largest protozoa can barely be seen as tiny specks in water with the unaided eye. Most of them are large enough so that their structures can be easily studied with the compound microscope, though some are small enough to make such study difficult. They are almost all big, though, as compared with bacteria, and giants in comparison with the viruses. Their endless variety and movement make them favorite subjects for anyone who, by peering through a microscope, wants to look into the strange world of subvisible life.

Some protozoa live together as colonies, each cell still a complete and potentially independent organism, but bound together with its brothers to live a communal life. It is an easy step from such colonies of protozoa to the members of the next great animal phylum, the sponges (porifera).

SPONGES

The sponges are generally considered to lie outside of the main lines of animal evolution, to represent a group of collared protozoa that learned to live together as colonies sufficiently well so that they were able to build up into great congregations of cells held together by a mass of fibers, spicules and secreted jelly. The fibrous skeleton of one type of sponge is familiar to all of us. The sponge cells co-operate to form a supercellular organism in which water moves along regular channels through the pores so that food materials can be extracted and waste materials thrown out, and the cells may be specialized in various ways. Parts of a sponge will reproduce the whole animal, like cuttings from plants. Reproduction is usually by a ball-like group of undifferentiated cells called a "gemmule" which swims away from the parent sponge, but there is also a complicated form of sexual reproduction.

COELENTERATES

In almost all arrangements of the animal kingdom the third phylum, after the protozoa and the sponges, is that of the coelenterates—the jellyfish, corals and sea anemones. About five thousand species are known, all marine except for a few inconspicuous fresh-water forms. The coelenterates are definite, organized, multicellular animals. The individual cells are organized into tissues with distinctive functions, and one may recognize digestive, muscular, nervous, sensory and even skeletal systems, though respiratory, excretory and circulatory systems are lacking. Distinctive sex cells, with reproductive functions, are produced, but they do not form part of a reproductive system as in more complex animals.

Coelenterates show radial symmetry, with all of the different parts disposed within the circular form centering on the mouth opening, instead of the bilateral symmetry of most animals. There are two body types, the medusa form (jellyfish), which is free swimming; and the polyp form (sea anemone), which has a fixed position. Within the same species, there may be an alternation of generations, the fixed polyp producing free-swimming medusae which, through sexual reproduction, give rise to a generation of polyps again. The same animal (or the same kind of animal) may thus look very different at different stages in its life history.

The coelenterates also tend to form colonies. Coral reefs are examples of gigantic colonies of polyps formed by the skeletons of the innumerable individual animals. The Portuguese man-of-war—familiar to people who have done much swimming in tropical waters—is an example of another kind of colony, made up of several individual animals with different forms and functions. A whole collection of animals hangs under the bright, bladderlike cell—some individuals specialized as feelers and stingers, others for eating, others for protection, and still others for reproduction.

In this rapid scanning of the animal kingdom we cannot list all of the phyla—indeed, no two authors agree on exactly how many phyla to recognize. There are, in the sea especially, a number of very peculiar small groups of inconspicuous animals, different enough to be considered as distinct major divisions of the animal kingdom. These may represent leftovers of types that were abundant at some time in the long geological past, or they may represent types of organization that did not prove to be "successful" in the sense of leading to a great development of species, of individuals, or of achieving a key position in the economy of nature. Miss Libbie Hyman, in an authoritative review of the classification of animals, lists 22 major phyla. Of these only nine seem conspicuous enough or important enough for review here. The others, all marine except for a few fresh-water forms, may be very interesting to the student of morphology or evolution; but each phylum has at most a few hundred living representatives, and none is conspicuous enough even to have a widely recognized vernacular name.

NEMATODES

The next major phylum, skipping several of these obscure but very interesting groups, is that of the nematodes. These might also be considered to be obscure, in that they have attracted relatively little attention from most people, but they are incredibly numerous and apt to be important in any consideration of the general economy of nature.

The nematodes are worms. This word "worm" is used for a variety of different kinds of animals—anything that is long, round, wriggly, and too small to be called a snake. Thus, all sorts of insect larvae are worms, though they would perhaps more properly be called grubs, caterpillars, or maggots. The true worms, that is animals that keep the worm form all of their life, include two very different phyla, the nematodes and the annelids. The nematodes include the worms with a smooth, unsegmented body, for which biologists have attempted to establish the vernacular terms "roundworm" or "threadworm", while the annelids include the segmented worms, such as the common earthworm.

Something like 80,000 different kinds of nematodes have been described and probably only a small proportion of the existing kinds have been named. They are much more highly organized animals than the coelenterates: the digestive system is a complete tract, with both mouth and anus, and the sex cells are associated with a reproductive system, so that individual animals may have male or female characteristics. Most nematodes are very small, microscopic, and while many kinds live in salt and fresh water, others live in soil or other moist situations. A great many also are parasitic, living in plants, insects and vertebrates. Some of the human intestinal worms are nematodes. Filaria, forming a large group of vertebrate parasites, are also nematodes; they include the organism that causes "elephantiasis" in man.

ANNELIDS

It would perhaps be best to discuss the segmented worms, or annelids, next, even though in classifications they are usually placed quite separately from the nematodes. The annelids include the earthworms, the leeches and a variety of marine worms. They are considerably more complex organisms than the nematodes, with all of the main organ systems of the higher animals. Their chief characteristic is the *segmentation* of the body, which affects not only the outward form, but the internal arrangement of the organs.

Only about a tenth as many annelids (8,000) as nematodes have been described, but the earthworms, at least, are numerous enough as individuals to be very important in the economy of nature. Charles Darwin became interested in the abundance and activity of earthworms, and wrote a book about them in which he showed that they were primarily responsible for the maintenance of favorable soil conditions, at least in temperate latitudes.

Most kinds of annelids are marine. It is generally supposed that life arose in the sea, and certainly conditions in the sea seem more favorable to life than in any other major type of environment. Most animal phyla show their greatest development in the sea, including only occasional groups that have solved the problems of adaptation to fresh water. Groups that have succeeded in adapting themselves to terrestrial life are even more sporadic, and the vast bulk of terrestrial animals belong to only two phyla, the arthropods and the chordates (vertebrates). The nematodes and the annelids include many terrestrial forms, but these are still limited to damp situations (soil, or parasitism in other organisms). The next great phylum, the echinoderms, is purely marine.

ECHINODERMS

The echinoderms include the starfish, sea urchins, sea lilies and their relatives. The adult organisms show a radial symmetry, like the coelenterates, but it is generally considered that this symmetry is a secondary development, since larval forms are bilaterally symmetrical. Indeed, because of the structure of these larval forms, it is thought that the echinoderms may lie close to the main stem of the ancestry of the most complex animals, the vertebrates.

About 6,000 living species of echinoderms have been described. They form a distinctive group because of the radial symmetry (almost always with five basic segments), the horny skin (formed by intracellular spicules of lime) and the curious hydraulic apparatus that furnishes power for movement.

MOLLUSCS

The shellfish and their relatives—the snails, clams, slugs, squids and octopuses—make up the next phylum, that of the molluscs. These are mostly marine, but many forms have invaded fresh water and many snails have become completely adapted to terrestrial existence. The molluscs form a very large group, numerous both in species and individuals, with something like 80,000 described living kinds. They show a complex development of organ systems like that of higher animals, but they lack the jointed appendages of arthropods and chordates. The shell is of course the most obvious characteristic of the molluscs, though this is shared with another phylum of marine organisms, the brachiopods, which has not been described here because there are so few living species. And many molluscs, such as the slugs of the garden and the squids and octopuses of the sea, have no obvious shell. One mollusc, the giant squid, is the largest of all invertebrates, attaining a length of more than fifty feet.

ARTHROPODS

Two phyla remain to complete this review of the animals—the arthropods and the chordates. These include the dominant land animals of our geological time—the arthropods because of the insects, which are the most numerous of visible organisms both in number of kinds and in number of individuals; and the chordates because of the mammals, which gain a peculiar interest because we are one of them.

The arthropods include five main classes, the crustacea (crabs, shrimps, barnacles), the arachnids (spiders, scorpions, ticks), the onychophora (the few queer species of Peripatus that have survived into our time from the dim geological past), the myriapods (millipedes and centipedes) and the insects.

To us as mammals, the insects seem to belong to some topsy-turvy world almost outside the reach of our understanding. They have the skeleton on the outside of the body, the main nervous system below the digestive tract; they carry air to the cells of the body through a complicated system of tubes and use blood (body fluid) only for the transport of food materials. They may become involved in amazingly complex patterns of behavior, but these patterns seem to depend on some inherited fixed instinct, not in any way involving a learning process. And the insects are endlessly abundant, endlessly prolific, endlessly obtruding themselves on human consciousness either in the garden or the kitchen. We, in our snobbish pride, consider this to be the Age of Mammals; but even we sometimes stop to wonder whether these ubiquitous insects are not destined, after all, to inherit the earth.

CHORDATES

Biologists now speak of the chordates, instead of the vertebrates, because they have recognized the cousinly relationships of some marine organisms which, although they have no backbone, must definitely be included in the family circle. These non-vertebrate chordates, however, include only a few inconspicuous species of little concern in a broad survey of natural history. For our purposes we can still think of the chordates as composed of five main classes, the fish (sharks and rays should really be a separate class), the amphibians, the reptiles, the birds and the mammals.

The chordates, with 70,000 or so species, make a respectable showing in comparison with any phylum except the arthropods; and they include organisms that dominate, in size if not in numbers of individuals, in all of the major types of environments.

Mammals are the most complex, the most specialized, the most recently evolved of the classes of chordates. With pride, we place man at the tip of the mammalian line, adducing his brain size, his manual skill, his geologically recent intrusion on the landscape, as arguments for this position. With the mammals and with man we thus come to the end of the survey of the animal branch of organic nature, having reached an extreme of specialization comparable to that of the orchids and the daisies on the plant branch.

To discuss "higher" versus "lower" organisms may well, however, be considered a foolish diversion. After all, any organism that is alive today represents success in the solution of the never-ending, multiform problems of survival, the bacterium as much as the monkey or the orchid. No living organism can be considered as ancestral to any other living organism, and even to class one as "primitive" and the other as "specialized" may be very misleading, since both, if they are alive, have obviously attained "success."

Which brings us to the question of the history of life on this planet, the subject of the science of paleontology.

CHAPTER IV

The History of Organisms

THE Museum of Comparative Zoology, at Harvard, was a fine building for its day; but now it is perhaps best characterized as a firetrap. It houses the accumulated hoardings of many naturalists, an irreplaceable treasure of biological materials. Consequently every precaution is taken to be sure that its firetrap possibilities are not realized. Among other things, no one smokes in the building. Scientists smoke at least as much as other people, and the scientists at the M.C.Z., compelled by their habit, spend intervals on the museum steps all day long, even in the worst Massachusetts weather. This means that the mammalogists, herpetologists, entomologists, paleontologists, and members of all of the other tribal divisions, are forced by their nicotine addiction to come out of hiding at frequent intervals, and associate informally on the steps. Some, of course, remain aloof, wrapped in the abstractions of their tribal thoughts; but most are impelled to talk, sometimes personal gossip, sometimes about what they have just been doing or what they intend to do next year. The insect man learns something of the bird man's problems, the shell collector something about fossil reptiles; and the learning is absorbed painlessly, along with the hourly cigarette.

This all started because I remember one sleety morning when

I was huddled under the entry stairs, to get some protection from the Boston weather. I was joined by Teddy White, a friend belonging to the tribe of paleontologists. I had often watched him about the painstaking business of freeing a fossil bone from its matrix, of trying to sort out a collection of teeth, jaw fragments and vertebrae.

This morning Teddy was feeling rather low. He had, I suppose, been trying to reconstruct some prehistoric monster, with only a piece of the left hind leg on which to build his reconstruction.

"Paleontology," Teddy remarked, "is like trying to put together a jig-saw puzzle when you have only a few of the pieces and no idea of the pattern."

That was many years ago, but the phrase has always stuck in my mind. I think of it when I see the reconstruction of a dinosaur, when I look at a painting of someone's idea of an Eocene landscape, or when I read about the lines of evolution in fossil horses.

It takes a rather special sort of a person to make a paleontologist, since the science requires both a vivid imagination and a painstaking attention to detail. The combination has resulted in an amazing accomplishment, in the reconstruction of a considerable amount of biological history from fragmentary and obscure materials. Darwin, in the *Origin of Species* emphasized the "imperfection of the geological record." Truly, the record is very imperfect, and must always remain so; but I think that Darwin, if he should revisit us today, might be more impressed with the advance in paleontology than with that in any other field of biology.

Some background knowledge of this geological and paleontological history is necessary for the understanding of the living phenomena that we see about us today. Hence it may be well to include here a sketch of the subject.

THE AGE OF THE EARTH

The question, how old is the earth, has long interested both scientists and philosophers. Bishop Ussher settled the matter nicely for a while by the announcement, in 1654, of the results of his scriptural calculations, which demonstrated that the earth had been created at nine o'clock in the morning on October 26th, in the year 4004 B.C. This time limit pretty well controlled both historical and geological thought until the publication, by James Hutton in 1785, of a paper entitled "Theory of the earth, or an investigation of the laws observable in the composition, dissolution and restoration of land upon the globe". Hutton is the father of geology, but he also deserves a prominent place in any history of biology, since he was the first to give the biologists time in which to allow for the development of life. The line of ideas is direct, from Hutton to Lyell to Darwin; and in his earlier years, Darwin thought of himself more as a geologist than as a biologist.

Hutton argued that the present is the key to the past, and that the geological features of the earth must have been produced by processes that are now at work, operating through long periods of time. This became known as the doctrine of uniformitarianism, and it has long been universally accepted by scientists. Such a doctrine must be a basic credo of science because if, to explain some phenomenon or other, a process not now observable is postulated, the explanation is thereby removed from the field of science, since it is not subject to observation or experiment.

It required an immensity of time to account for the physiographic features of the earth by Hutton's doctrine, and many attempts have been made to find some measurement of this time. Until the twentieth century, only two methods of measurement seemed promising—the calculation of time required for the deposition of sediments, and the calculation of time required to account for the salinity of the ocean. Both were very imperfect methods. The difficulties of the sediment calculations are obvious, since there is no way of knowing what sediments have been lost, or at what rate sediments accumulated at different periods in the geological past.

The salinity of the ocean is also an uncertain measure, since much of the salt of the sea is constantly involved in cycles whereby it is withdrawn in land deposits and eroded again into the sea, so that even if one could arrive at an estimate of the amount of salt added to the sea yearly by our present rivers, one still would not know how much of this had been removed from the sea, and how much derived from rocks not previously eroded.

Geologists currently consider the decomposition of radioactive minerals to be their most accurate measure of time. For instance, the rate at which uranium disintegrates into lead is known with great accuracy; and the rate seems to be constant, regardless of circumstances. The ratio of uranium to lead in a given mineral, then, should give an accurate measure of the age of that mineral. Enough uranium-bearing strata have been found to give estimates for several geological periods, and these estimates fit well with other types of calculations, and with the immensity of time that is required for geological and biological processes to have reached their present conditions.

The oldest rocks have an age, by this method, of about two billion years, which sets a minimum for the possible age of the earth. The first adequate and clearly recognizable fossil remains of life date from the Cambrian period which, by the uranium clock, was five hundred million years ago. For the first billion and a half years of geological history, during which life reached a fairly high degree of complexity, as witnessed by shell-bearing molluscs, we have no details of the record.

These immensities of time pass easy comprehension, and perhaps the best we can do is quote the analogy of Sir James Jeans: "Let the height of the Woolworth building represent geologic time. We may then lay a nickel on its tower to represent the time of human existence. A thin sheet of paper on this will represent all historic time!"

THE DIVISIONS OF GEOLOGICAL TIME

The geologists of the world have devoted a great deal of thought to the development of a table of geological time. This is based on two general principles—the law of superposition and the law of faunal succession. The first recognizes that in any given region, the sedimentary deposits form a time sequence, with the oldest at the bottom; this, of course, may be upset by the movements of the earth's crust involved in mountain building, but such deformations can be recognized. The second principle means that a given type of sedimentary rock of a given age includes among its fossils the members of a particular time fauna. The characteristics of the fossil species change slowly from age to age with evolutionary development. When fossils of a given type are found in a sedimentary rock in one country, fossils of the same type in rocks in other countries give a strong indication that the rocks represent the same geological period.

The record, of course, is very incomplete in any one place. Sediments have been deposited, raised, eroded again into new sediments repeatedly, so that the pieces of the record left undisturbed from any given era are scattered widely over the surface of the earth; and there are long periods, indicated by abrupt changes in the types of fossils, for which no sediments whatever have been found.

Slowly, however, the geologists have put together a table of geologic events which is reasonable enough to have received fairly universal acceptance—an acid test, since scientists do not readily agree with one another. They have divided the history of the earth into major eras, separated by great gaps in the record presumably caused by periods of mountain uplift and consequent extensive erosion; these eras have in turn been divided into periods, the periods into epochs and stages.

This is comparable to the methods of the archeologists and

historians in making a table for the events of human pre-history and history. In history the eras are represented by civilizations, separated by gaps caused by barbarian invasions or other destructive processes. Thus in the northern Mediterranean we have the very definite Minoan civilization, separated from the Graeco-Roman by a gap in the record, this in turn separated from contemporary Western civilization by the gap of the "dark ages". If these were eras, the periods, say in the Graeco-Roman, might be Athenian, Roman Republic, Early Empire, and so forth; the epochs might be Periclean, Alexandrian, Augustan, to take three examples at random.

The geologists recognize five major eras, preceded by an unknown stretch of "Azoic time". These eras are the Archeozoic, the Proterozoic, the Paleozoic, the Mesozoic and the Cenozoic. The rocks of the Archeozoic and Proterozoic are practically devoid of fossils, perhaps partly because the early forms of life did not have hard shells that would readily fossilize, partly also because the transformations of these rocks through subsequent heat and pressure would have destroyed the fossil record. There are impressions in the rocks that fairly surely represent calcareous algae and sponges, and there are trails and burrows that were probably left by some kind of worm-like creatures; that is about all. There are, however, deposits of carbon in these rocks which are inexplicable except as the last trace of some form of animal or plant; and, from the extent of the deposits, this life must have been abundant.

The commonly recognized periods of the three remaining eras are listed in the accompanying table, together with some indication of the sort of organisms that are known to characterize the different periods. In geologic history, as in human history, the record tends to be clearer as contemporary time is approached. But in both cases, this tendency is far from regular, due to the accidents of the preservation of materials. We know a great deal about Egyptian history because the climate of the Nile valley is very favorable for the

THE PERIODS OF BIOLOGICAL HISTORY

Era	Period	Time*	Characteristics
Paleozoic	Cambrian	550	Marine fossils: molluscs, crustacea, algae.
	Ordovician	475	Probable first vertebrates: jawless fish.
	Silurian	400	Probable first insects and land plants.
	Devonian	350	Amphibians; forests.
	Mississippian	300	The "carboniferous" period: vast forested swamps (ferns); insects, amphibians.
	Pennsylvanian	250	
	Permian	210	Extinction of many animal types; early reptiles.
Mesozoic	Triassic	200	Many reptile groups; coniferous plants.
	Jurassic	150	Toothed birds, archaic mammals; dicotyledonous seed plants.
	Cretaceous	120	Extinction of dinosaurs; monocotyledonous plants.
Cenozoic	Paleocene	65	Climax of archaic mammals.
	Eocene	60	Forests of modern plants; rise of modern mammals.
	Oligocene	40	First rats and first anthropoids.
	Miocene	35	Grasslands; climax of mammals.
	Pliocene	25	Evolution of man.
	Pleistocene	1	Periodic glaciation; extinction of many mammal types.

* Estimated time in millions of years since the beginning of each particular period.

preservation of records, because the habits of the people left records liable to be preserved, and because, with the Rosetta stone, we learned how to read them. We know very much less about the Minoan civilization that was flourishing at the same time because the records are far less complete, and because we cannot read such records as have survived. A civilization that built with wood might have left no trace at all. The difficulties and hazards of human history and pre-history are multiplied thousands of times in paleontology.

PALEOZOIC TIME

Fossils from the Cambrian are abundant, but they are all in marine deposits. We can thus get a fairly detailed idea of what life in the Cambrian seas was like, though we have no record of what existed on land. There was a great variety of life in these seas, including representatives of almost all of the main phyla of animals. Most of the fossils, of course, are of animals with hard shells, readily preserved, and the commonest kinds are trilobites, queer heavily armored crustacea. Most of these were small, but one kind, 18 inches long, is the largest animal known from the Cambrian.

On the flanks of Mt. Wapta, in the Rockies of British Columbia, a marvelous fossil record of the Cambrian has been found. This is in black shale, and in the planes of the shale are beautiful impressions of soft-bodied creatures that lived in the Cambrian seas—jellyfish, sponges, seaweeds, a variety of bristled annelid worms, and delicate shrimps. These are all complex organisms, and must represent the results of evolution through an immensity of pre-paleontological time. No signs of vertebrates, however, have been found in the Cambrian.

The paleontologists have many reasons for believing that the first vertebrates developed in fresh water. There are no fossils known from the Cambrian that might have been formed in fresh water.

The same is true of the Ordovician except for one deposit that may have been formed at a river estuary, and that contains bits of skin armor that probably belonged to a jawless fresh-water fish. Silurian sediments again are almost all marine, but some that may be from fresh water or from river estuaries contain abundant remains of many kinds of jawless fish, animals somewhat akin to our lampreys. There must also have been other types of vertebrates by this period, because many types of fish and some fragments of amphibia are known from the Devonian.

The first abundant remains of land plants also date from the Devonian, including representatives of all of the main plant groups, except the seed plants. Some of these fossils are stumps and trunks of large trees, in part clearly belonging to the fern phylum. The relationship of others is uncertain. One, Nematophyton, is constructed like an alga; but since some of the trunks were three feet in diameter, the plant must have been totally unlike any alga that we can imagine today. With these plant fossils, remains of spiders and mites have been found, the first evidence of terrestrial arthropods.

The Carboniferous period, often split into the "Mississippian" and "Pennsylvanian" by American paleontologists, left an abundant history in the great coal beds. These coal beds are thought to have been produced by vast forested swamps, the trees including something called "seed ferns" and "scale trees" as well as giant ferns and giant horsetails. The biota of the coal-bed period was very rich. It seems queer to us chiefly because neither flowering plants nor warm-blooded animals had appeared. Insects, including flying forms, were abundant, but of orders long since extinct; there were many amphibians and some reptiles, also of types that have become extinct.

MESOZOIC TIME

The Mesozoic era was the "Age of Reptiles", of the dinosaurs, the flying pterosaurs, the stream-lined, marine ichthyosaurs—of a host of reptilian types invading all of the major environments. These all disappeared at the end of the Cretaceous, and the reptiles that survived into the Cenozoic era, the turtles, lizards, snakes and crocodiles, form an inconspicuous assemblage beside this Mesozoic profusion. The possible causes of the reptilian catastrophe have been much debated by paleontologists. The extinction was probably a slow process, however abrupt it seems looking back over the fossil record, and it was probably related to major climatic changes associated with the beginning of the elevation of the Rockies and other mountain chains throughout the world. It is generally thought that the highly specialized, the bizarre and conspicuous reptiles, failed to produce the adaptations necessary to meet these changes, so that the earth was inherited by the meek, the inconspicuous, unspecialized primitive mammals that have left their first traces among the fossils of the Cretaceous.

CENOZOIC TIME

The history of the Cenozoic is the history of the development of the dominant land biota of the world today, of flowering plants, of contemporary orders of insects, of birds and mammals.

The Cenozoic era is often also called the "Tertiary", particularly by English writers. This is a relic from the classification of Giovanni Arduino, an Italian who dedicated much study to the rocks of the southern Alps. He proposed, in 1759, to group these rocks into four series which he called primary, secondary, tertiary and volcanic. The first series included the schists that form the core of the mountains, the second the hard sedimentary rocks on the mountain flanks, and the third the less hardened sedimentary rocks of the foothills. This classification was logical for the Alps, but it did not fit other situations and the terms "primary" and "secondary" were gradually dropped, though "tertiary" has persisted with a restricted meaning, either as a direct synonym of Cenozoic, or as a term for all of the Cenozoic except the last period, the Pleistocene.

The divisions of the Cenozoic at present generally recognized were first worked out by the great English geologist, Sir Charles Lyell. He and a French shell student, Deshayes, based the divisions on the shells in a series of deposits in the region of Paris which was inundated periodically by the Cenozoic seas. Many of the shells left as fossils after these inundations were of species still living, the percentage of living species depending on the age of the deposit. Thus the earliest period, the Paleocene, included no living species; 1 to 5 per cent of the species found in the next period, the Eocene, were the same as species still alive; in the Oligocene, 10 to 15 per cent were living species; in the Miocene, 20 to 40 per cent; in the Pliocene, 50 to 90 per cent; in the Pleistocene, 90 to 100 per cent.

THE PLEISTOCENE AND GLACIATION

The Pleistocene represents the end of geological history. When the Pleistocene ends and "recent" or "modern" or "contemporary" time begins is a rather arbitrary matter of definition. The entire Pleistocene probably covered a million years or so, and during this period, the lands of the northern hemisphere were repeatedly covered by great glaciers. The last of these glaciers reached its greatest extent only a few seconds ago, by geological time, and the beginning of the retreat of this glaciation, still represented by the icecaps of Greenland and Antarctica, is perhaps the most logical event with which to close geological history. This retreat probably began some 25,000 or 30,000 years ago.

The Pleistocene, as the most recent of geological periods, is also the best known, since the relics of its history have been least destroyed by the lapse of time. The outstanding characteristic of the Pleistocene was the repeated glaciation of Europe and North America. Tracks of the glaciers are obvious in many settled parts of Europe and North America, and during the seventeenth and eighteenth centuries these "diluvial" marks were attributed to Noah's flood. It is perhaps natural that the Swiss should be the first to recognize the glacial origin of these diluvial remains, and the great Swiss naturalist, Louis Agassiz, was largely responsible for the general acceptance of the glacial theory.

The geologic signs of glaciation are very distinctive, and all of the various phenomena can be observed with contemporary glaciers. Rocks are smoothed, polished and scratched in characteristic ways, and great boulders are often carried far from their parent formations in the ice stream, forming glacial erratics. A great accumulation of debris forms at the foot of the glacier, making a terminal moraine. These moraines may cut across normal drainage channels and, along with the direct eroding action of the glacier, give rise to the numerous lakes found in glaciated regions. All of these signs, of course, become less clear as one goes back in geological time, so that the history of glaciation cannot be traced with great accuracy in older periods; there is considerable evidence, however, of glaciation at various remote times, so that the phenomena of the Pleistocene are not unique.

It is now generally recognized that during the million years or so of the Pleistocene there were four major periods of glaciation in the northern hemisphere. At their maximum extent, these glaciers were vast, reaching down to the Ohio and Missouri rivers in North America, and to northern Germany, Belgium and the British Isles in Europe. There were smaller icecaps at the tip of South America, in Australia and New Zealand, and on high mountains in all parts of the world.

It is difficult to imagine all of the effects that this immense accumulation of ice must have had on the physical and biological conditions on the earth. The thickness of the ice is of course not known, but it must have been 5,000 or more feet thick at the center of dispersal, and perhaps even 10,000 feet thick. The present ice cap of Greenland is 8,800 feet thick near its center, and averages 4,500 feet thick over a large area. The ten million or so square miles of Pleistocene ice sheets would thus represent many millions of cubic miles of water withdrawn from the oceans, which must have resulted in a marked lowering of sea level, variously estimated at from 150 to 300 feet. During interglacial stages, the water would have returned to its present level, or perhaps even higher. The weight of this ice accumulation on the continents also seems to have resulted in very considerable changes in level and slope. The effect on air temperature can only be guessed; the temperate zones were surely much colder, but the effect on climate in equatorial regions may not have been marked.

The possible cause of Pleistocene glaciation is an open field for speculation, in which there is no general agreement. Probably there were several causes, one perhaps being continental uplift, which would increase the areas of cool highlands in temperate latitudes. Changes in atmospheric composition have been suggested, such as a lowering of the content of water vapor or carbon dioxide. A small change in the carbon-dioxide content of the air would change greatly its heat-absorbing capacity. Long-term variation in solar radiation might be the ultimate cause starting trends toward glaciation or thawing. It seems possible that relatively slight changes might initiate these trends. Ice represents a storage of coldness, so that once it has started to form and to persist from one cold season to the next, there would be a cumulative effect.

The Pleistocene glaciation, so recent in geological time, certainly serves to remind us that the climate of the earth, as we observe it today, is not necessarily constant. Long climatic changes may, in fact, have been an important factor in controlling the course of organic evolution, though a factor very difficult to evaluate. One of the characteristics of the Pleistocene, for instance, was the great reduction in the number of mammal types. Presumably all of the mammals that are alive today existed all through the Pleistocene, and in addition there were a great variety of mammals that became extinct during the period. This extinction cannot clearly be related to the glaciation, since many seem to have died out only very recently, after surviving all of the vicissitudes of the Ice Age. Such include the mammoths and other elephant relatives, and the horses, camels, and giant ground sloths of America. Man started to exert his effect on the rest of nature during this period, but it is hardly credible that he, directly, could have been responsible for the extinction of all of these other mammal types. We know that he lived together with many of them, as witnessed, for instance, by the European cave drawings of mammoths.

Man is a Pleistocene phenomenon, since all of the clearly human fossils date from this period. He must, however, have been developing for some time, and anthropoid fossils are known from as far back as the Oligicene. The history of man is thus bound up with the glaciers: but this history lies outside of the area of biology that we have started to explore.

CHAPTER V

Reproduction

THE fossil record, however imperfect, thus gives a clear outline of the slow development of life on this planet, from the now
formless accumulations of carbon in the Pre-Cambrian rocks to
the extraordinarily diverse types of organisms that populate the contemporary scene. As naturalists, we are concerned primarily with
the events of this contemporary scene, which the geologist labels
"Recent", but to understand these events we must constantly refer
back to the records of their historical development. Where the records are incomplete (as is usually the case) we are forced to fill in
the account with guesses, taking care to make our guesses conform
with the known fragments of the record.

All of the evidence indicates that life as we know it on this
earth has had a common origin and a continuous history. Individuals die, species become extinct, whole orders and classes and phyla
of organisms may flourish, decline and disappear, but they are replaced by new types and thus the germ plasm goes on forever, the
sex cells of each generation passing on the vital impulse of life to
the next. Each generation almost exactly copies the last, but it
sometimes differs slightly, sometimes varies and thus through the
eons of time, gives rise to the apparent diversity of organic form.

The continuity and the diversity depend on the process that we call reproduction.

Reproduction is a fundamental property of life. A given organism, any kind of organism, has the potentiality of producing other organisms of its same kind. This is an extraordinary business which seems to put living processes quite outside the scope of the laws that govern other aspects of the physico-chemical universe.

Reproduction is not only a fundamental property of life, it seems to be the fundamental purpose of life. We are, of course, headed for trouble when we get mixed up with purposes, but the concept of purpose is very important in human thinking, and we have no means of examining nature without using the instrument of human thought. With any process or any thing we are always asking what it is, how it works, and why it exists. There are various levels of answers to all of these questions, but with the why it seems never possible, with organic nature, to get below the apparent purpose of self perpetuation. All life processes seem dedicated to the final purpose of reproducing the individual, of perpetuating and increasing the species, of insuring the continuity of the life process itself.

THE CONTINUITY OF THE LIFE PROCESS

Biologists have come to think of all living things as representing a continuum, as diverse manifestations of an uninterrupted transfer of the living material from one individual to another, from one species to another, throughout the long history of life on this planet. This concept grew very slowly, though it is now so firmly rooted in the biological mind as to be almost subconscious. The idea of the essential continuity of life processes was not widely realized until after the publication of Charles Darwin's *Origin of Species* in 1859, and the idea of the possible "spontaneous generation" of life was not finally abandoned until the famous experiments of Louis Pasteur a few years later.

It is perhaps too strong a statement to say that the idea of the possible spontaneous generation of life was finally abandoned after Pasteur. An idea like that is hard to down. The ancients, of course, believed that flies arose from putrefying matter, and in the Middle Ages it was quite commonly believed that rats could be generated from bran and old rags. An Italian, Francesco Redi, who lived in the seventeenth century, first showed that the maggots on meat arose from eggs laid there by flies, but even Redi believed in the spontaneous generation of intestinal worms. The discovery of protozoa and bacteria, with their obscure methods of reproduction and dissemination, quite naturally gave a new impulse to the idea of spontaneous generation which persisted until the ingenious Pasteur showed that even bacteria had to come from other bacteria. With the discovery of viruses, and the increasing realization of the tenuousness of the difference between viruses, which are certainly alive, and some enzymes, which are certainly "mere chemicals", the spontaneous generation of life has again become an open field for speculation—though the speculation is more cautious, clothed with a more obscure vocabulary than in the early days.

We assume, of course, that life must have started somewhere, sometime; and if it started once, it may have started many times. The conditions necessary for the shift from the non-living to the living state may have been some special sort of environment that existed only in the dim geological past of this planet; or life may have arrived on the earth riding a meteor from outer space; or it may somehow arise continuously under our noses by some process that we have as yet overlooked.

The biochemists, working with enzymes and viruses, have the more or less explicit hope of some day achieving a synthesis of "living" material in the laboratory. From their point of view, it is essential to postulate that this is possible, and they have made great progress in analyzing the components of the simplest known living systems. From the point of view of natural history, however, it is most fruitful to postulate a single origin for life as we know it, and all of the evidence that has so far been collected tends to support this presumption.

THE CELLULAR BASIS OF REPRODUCTION

Reproduction, viewed broadly, is a vast subject, the concern of many different sciences. And it remains essentially a mysterious process, despite the multitudes of facts, observations and theories that have been accumulated by these sciences. The structure of the germ cells has been examined in great detail by the cytologists and biochemists. The development of the germ cells, whereby the single original egg multiplies and gradually takes the form of a definite kind of an individual, forms the subject of embryology. The study of inheritance, the tracing of differences among individuals to differences arising in the germ cells, constitutes the special science of genetics. The physiologists, studying the ways in which life processes work, become involved with many aspects of reproduction.

From the point of view of natural history, we are concerned primarily with the behavior patterns that surround the reproductive process, and with the potential rates of reproduction of different kinds of organisms under different circumstances. Methods and rates of reproduction will be of constant concern when we come to examine the relationships between different kinds of organisms, and between particular kinds of organisms and the physical environment.

We customarily consider that there are two basic kinds of reproduction, sexual and asexual. Sexual reproduction, whereby the new individual arises from the union of two germs, from two different parent individuals, or perhaps from the same parent individual, is such a widespread phenomenon that it must be considered a very fundamental attribute of the life process. Sexual reproduction has been presumed to be absent in only two major groups of organisms, the viruses and the bacteria, but even in these groups there is indirect evidence of some kind of a fusing, sex-like process occasionally occurring.

It is impossible to discuss reproduction without getting back to the cell, that common structural denominator of all organized life. Single-celled organisms (bacteria, protozoa, algae) may reproduce by simply splitting in two, a process perhaps comparable to growth in complex organisms insofar as growth is the result of the continuing division and multiplication of the separate cells that go to make up the organism.

This process of cell division, by "simply" splitting in two, is in reality far from simple. Any cell is a very complex affair, and this complex organization must be continued through the process of cell division. The organization of the cell is apparently controlled by a particular part, called the nucleus, and by particular complex molecular systems within the nucleus called chromosomes. The chromosomes can be seen with certain staining techniques, and it has been found that during cell division each chromosome splits so that an equal part of this basic organizing material is carried into each new cell. This process, called mitosis, is described in all biological textbooks (or in any encyclopedia) and is worth looking up.

The nuclear material, carefully passed on from cell to cell, somehow controls the whole organization of the cell, which means the whole organism if the organism consists of a single cell. But even in multicellular organisms this nuclear material controls the organization of the whole individual, since each such individual starts out from a single germ cell. In a complex organism the various kinds of cells become very highly specialized for different sorts of functions—for carrying nerve impulses, for contracting to cause muscular movement, for special chemical operations in the processes of digestion, respiration, excretion, for support as in bone or for protection as in skin, and so forth—but each such organism carries along also a supply of the original unspecialized cells, the germ plasm.

Actually, it is easiest to think that, potentially, any cell of a complex organism has the possibility of reproducing the whole organism, but that this potentiality has been gradually atrophied by the increasing specialization of the different cells. Any cell from a sponge can reproduce the whole sponge; a piece of an earthworm may grow into a whole earthworm; in some very complex plants a piece of leaf, stem or root may grow into a whole new plant with all of its different kinds of cells, showing that the cells of the leaf, stem or root have retained the ability to form all of the other kinds of cells. This subject of the regeneration of parts of an organism, or of the formation of a whole organism from a part, has been studied in great detail by experimental biologists, and forms a fascinating type of investigation. The scar tissue that forms in man after a wound is a final relic of this "totipotency" of his very specialized cells.

Asexual reproduction is common only among relatively simple animals. In the protozoa a single individual may split and thus become two. In the sponges, coelenterates and other less specialized phyla of multicellular animals, mechanisms for asexual reproduction may be present. A sponge for instance may produce small ball-like gemmules by budding, which swim off to form new sponges. In the coelenterates, the polyp form (hydra, anemone) may split off medusae (jellyfish form) by a process of budding; but in this case, the jellyfish reproduce by a sexual process, giving rise to new polyps, so that there is an alternation of sexual and asexual generations. Such an alternation forms the basic pattern of reproduction in plants, but it is not easily understood without some explanation of the cellular mechanism of the sexual process.

The sexual process involves, essentially, the fusion of two cells from two different parents to form a new individual. Now the organization of the cell, and hence of the whole individual made up of cells, depends on material in the cell nucleus called chromatin which, at the time of cell division, becomes organized into definite bodies called chromosomes. The cells of each kind of organism have a definite number of chromosomes, sometimes only a very few, sometimes many; man, for instance, has forty-eight; the fruit fly, four.

Obviously if ordinary cells fused, each carrying its complement of chromosomes, the thing would not work unless there were some mechanism whereby each chromosome would also fuse into its counterpart. But it doesn't happen that way. Instead, germ cells go through a special kind of cell division called the "reduction division" in which the chromosomes, instead of splitting, go half to one daughter cell and half to the other. The germ cells of a particular organism, in other words, have only half as many chromosomes as the other cells of the body, so that when two germ cells fuse, the proper number of the chromosomes for the species is restored, with half of the number coming from each parent. The normal complement of chromosomes is called the diploid number; the complement of the germ cells, the haploid number.

The total process is, of course, more complex than I have described here, and various stages and variations of the process have accumulated an involved vocabulary, which must be familiar to every working biologist but which seems hardly necessary for the nonprofessional. One of the most interesting subsidiary points is that all through the animal and plant kingdoms two kinds of germ cells can be distinguished. Each has the haploid number of chromosomes, but one is small and active, practically reduced to the nuclear material, while the other is large, relatively inactive, with a great deal of cytoplasmic material besides the essential chromatin. The first is the male element, the sperm, the microgamete; the other the female, the egg, the ovum, the macrogamete. "Gamete" (from the Greek *gamos,* marriage; *gamete,* wife; *gametes,* husband) is the formal word for a germ cell and to avoid the connotations of talking about "male" and "female" among such organisms as protozoa or mosses, one can refer to the two as microgametes and macrogametes. When the two have fused, the result is a "zygote" (Greek, yoked together), if it seems advisable not to refer to the thing as a fertilized egg.

The production of microgametes and macrogametes, and the contrival of circumstances to bring the two together, is subject to endless variation among the different groups of the animal and plant kingdoms. The subject is so very important from the point of view of natural history, of ecology, of the environmental relationships of organisms, that it may be advisable to pause long enough to give some examples.

Sexual behavior takes a wide variety of forms even among the single-celled organisms, protozoa and algae. Here, of course, the whole animal becomes a gamete. In Paramecium, a protozoan that is easily maintained in the laboratory and hence subject to much study, a phenomenon called "conjugation" occurs. This animal normally reproduces by fission, but at certain times (usually when conditions in the culture medium become adverse) individuals will be seen to pair off, their two bodies partially fusing for a time, and exchanging nuclear material; the two animals then separate and each continues to reproduce by the normal method of fission. Paramecia about to conjugate look just like ordinary Paramecia, but apparently any given strain has two different "mating types" which conjugate with each other, but not with members of the same type.

In many protozoa, microgametes and macrogametes are clearly differentiated, and the sexual process involves complete fusion of the two gametes. The exchange of nuclear material described for Paramecium is, in fact, a rare phenomenon, characteristic of that particular group of organisms. Plasmodium, the parasitic protozoan that causes malaria in man, will serve as an example of more obviously complex behavior.

Plasmodium is a very specialized parasite which lives within the red blood cells of particular vertebrate hosts, and the three or four species that parasitize man may cause, through the destruction of these cells, very severe disease. The cellular parasites normally divide by fission, but at certain times different appearing parasites can be found in some of the blood cells, sexual forms called microgametocytes and macrogametocytes. These never reproduce further by fission, and finally die unless the bit of blood that they inhabit happens to be sucked up by a mosquito.

In the stomach of a mosquito, these cells undergo a rapid and drastic modification, the microgametocytes split into four to eight very active microgametes which struggle out of the blood cell and swim actively about searching for the macrogametes, which have also freed themselves from their blood cells. A single microgamete fuses with and fertilizes a macrogamete, which thereby becomes a zygote. The malaria people, of course, have given special names to all of these things, and call their zygote an "oökinete". This, by whatever name, penetrates the stomach wall of the mosquito and forms a cyst there, dividing repeatedly to form a large number of individuals called sporozoites. These eventually rupture the cyst, and move about in the body of the mosquito until they come to rest in the salivary glands. There they wait to get into the blood stream of the next animal that the mosquito bites. If this animal happens to be the proper kind (man, in the case of human Plasmodia) they penetrate the blood cells and go about the business of asexual reproduction.

Sexual reproduction has thus got involved with all sorts of things. It takes place in a completely different environment from the asexual reproduction that serves to maintain the population of the species—the one in the mosquito stomach, the other in vertebrate blood—and it becomes tied up with the problem of the dispersal of the organism. The occurrence of the sexual process in the mosquito insures that the species will be transferred to new hosts, and reproduction is thus bound up intimately with the general problems of the environmental relations of the animal.

In multicellular organisms, the gametes are typically produced by a special tissue of germ cells which is recognizably distinct early in the history of the individual, and these germ cells are usually associated with a special system of reproductive organs. In most marine animals, the gametes, sperm or eggs, are simply liberated into the water in enormous numbers, fertilization depending on contact between the gametes in open water. Such marine animals are also often sedentary, occurring in colonies or aggregations, and the release of gametes from various individuals may be synchronized by environmental stimuli. Individual animals may release gametes of only one type, or both types of gametes may be produced by one individual, which would then be called hermaphroditic. In the case of active marine animals, such as fish, the sperm may be released directly over the eggs or in the immediate vicinity of the female.

MODIFICATIONS OF THE SEXUAL PROCESS

Adaptation to terrestrial habitats required the use of some more direct means of insuring contact between the gametes of a given species, and the most common solution of the problem among animals has been internal fertilization of the eggs in the parent female through copulation with the male. All sorts of elaborate behavior patterns have come to be associated with the process, to insure that individuals of a given species will be in proximity at the proper time.

Among plants, an alternation of sexual and asexual generations in a given species is the rule. In the algae and fungi a variety of complicated reproduction sequences may be found. In the mosses, ferns and seed plants, there is a clear distinction between the game- tophyte (gamete bearing) and sporophyte (spore bearing) genera- tions. The obvious plant among the mosses is gametophyte; it pro- duces ova and sperm, and the ova are fertilized by sperm that reach them swimming through a film of water from rain or dew. The fertilized ovum grows into a sporophyte plant, living as a sort of parasite on the gametophyte plant; this, when mature, produces an enormous number of spores which serve to insure the dispersal of the species.

Among ferns, the sporophyte is the obvious plant, and the little brown packets of asexual spores on the under side of the leaves are familiar objects. These spores, if they chance to land in a favor- able environment, produce a tiny gametophyte; again, as in the mosses, fertilization is dependent on the microgametes reaching the macrogametes by swimming through a film of water; after fertiliza- tion, the zygote starts to develop, producing a sporophyte, an obvi- ous fern plant; the gametophyte, its function fulfilled, shrivels away. Thus in both the mosses and the ferns, the multiplication and dis- semination of the species depends on asexual spores which always, however, produce sexual plants—the conspicuous plant in the case of the mosses, the inconspicuous one in the ferns.

In seed plants, the gametophyte generation has become so re- duced that it would not be recognizable if it were not for analogies with the more obvious alternation of generations in mosses and ferns. A pollen grain of a seed plant is not the exact equivalent of a sperm cell of an animal; rather it is a sort of telescoped game- tophyte consisting essentially of one vegetative cell and two germ cells. Similarly, a sort of abortive gametophyte is formed within the ovule of the flower, one of whose cells forms the ovum that is even- tually fertilized when a suitable pollen grain lands on the stigma of the pistil whence its sperm can reach the ovum through a pollen tube that grows down from the pollen grain. The fertilized ovum then grows into an embryonic plant enclosed with stored food material to form the seed.

Thus the sexual process, essentially the same in simple aquatic plants and animals (though varied enough in details of mechanism from group to group among either plants or animals), has followed basically different lines of development in the more complex organisms of the two kingdoms, to achieve mechanisms adapted to the exigencies of terrestrial existence.

THE PARTICULATE BASIS OF INHERITANCE

The universality of the sexual process is closely related to another peculiar biological phenomenon, the particulate basis of inheritance. We all know that offspring resemble parents, and also that offspring differ from parents. But the analysis of the resemblances and the differences eluded biological minds until a Dutchman, Hugo de Vries, discovered in 1900 a paper published by an Austrian monk, Gregor Mendel, in 1865.

Mendel must have been a very interesting person. He was born of peasant parents and probably entered the monastic life because it was his only possible chance to study. He devoted himself to mathematics and natural history (a rare and excellent combination) and became schoolmaster and eventually prelate of the Augustine monastery at Brünn. His leisure time was spent in experimenting with the inheritance of different characters of peas that he grew in his garden, and he published the results of these studies in two carefully prepared papers that appeared in the journal of the local natural science society—to be completely ignored by the great world of science until long after the monk was dead.

Essentially, Mendel discovered that the characteristics of a given organism are inherited as units. He discovered this because his first experiments were with crosses between varieties of peas that differed in only one character—dwarf versus tall, smooth versus wrinkled, yellow versus green and so forth. When he proceeded to try crosses between varieties that differed in two or three characters, he was able to determine that the inheritance of each character was quite independent. This discovery would have been difficult, even with Mendel's ability at mathematical analysis, if he had started his experiments with crosses between strains that differed in several characters. Mendel's work is a fine example of the importance, in experimenting, of attacking the simplest possible situation first and then proceeding to the more complex situation.

The modern science of genetics was produced by the synthesis of Mendel's discovery of the unit nature of inheritance with the cytological discoveries of the phenomena of mitosis and maturation of the germ cells, mentioned earlier in this chapter. Genetics has developed into a vast field of knowledge, which has been turned to very practical account by agriculturists, and which has been widely publicized by eugenists (or whatever proponents of eugenics are called) as support for their alarm about the future of mankind. The word "gene", coined by a great Danish biologist, Wilhelm Johannsen, has become part of the common language; and we all now lay the blame for our defects on our genes, or claim credit for the brilliance of our children because of the genes we have passed on to them.

Modern genetics certainly represents a respectable achievement of the human mind, compounded of bold speculation checked by ingenious experimentation, and geneticists have learned a tremendous lot about the physical basis of heredity. The genes have been found to represent specific points along the chromosomes of the cell nucleus, which control the organization of the growth of different aspects of the whole organism. The chromosomes of the fruit fly, Drosophila, have been mapped in great detail, so that hundreds of points are known which control definite traits of normal or variant flies. The history of these chromosomes has been traced through the complexities of mitosis and maturation, and certain types of variations found to be associated with certain types of acci-

dent that can take place in those processes. The redistribution of
genes in each generation through the operation of the sexual process
is well understood, and a beginning has been made in the study of
gene distributions in natural populations.

Genetics was for a long time the object of suspicion by other
types of biologists, because the geneticists were involved almost exclusively with laboratory problems which could hardly be interpreted in terms of wild populations. Now, however, the geneticists
have gone into the field where they are developing one of the active
growing points of natural history, the study of variation in natural
populations.

REPRODUCTION, INHERITANCE AND EVOLUTION

The whole weight of genetic knowledge goes to show that the
inherited potentialities of the organism are governed by these genes,
by the determining organization of the material of the chromosomes
of the germ plasm. The development of these potentialities may be
influenced by the environment, but it seems that the environment
cannot in turn influence the chromosomal organization directly.
The environment can only act by allowing for the survival of favorable gene combinations and by eliminating unfavorable combinations—the process known as natural selection. The general environment does not induce new genetic potentialities—acquired characters are not inherited. New hereditary potentialities apparently must
arise through chemical accidents within the chromosomes—through
mutations. The rate of such accidents, of mutations, has been cal-
culated for various species and situations, and found to be very
slow, though not too slow to meet the statistical requirements of
change within the geological time available. These rates can be influenced by certain external factors, notably bombardment with
radiations, but the resulting hereditary variations are of a random
sort, mostly monstrosities.

But while we know a great deal about the way certain characters are inherited, and about the way new characters may appear, the relation of this knowledge to the process of evolution in organic nature is still obscure. One influential group of biologists believes that there are essentially two sorts of evolution: an evolution of specific characters, of details of structure and function such as the geneticists have been able to study in the laboratory; and some other sort of process governing the development of groups, of genera, families, orders, phyla of organisms.

I find it difficult to believe that there is a qualitative difference between microevolution and macroevolution because of the difficulty of drawing sharp lines between the various categories of animals. Decision as to whether a given population represents a subspecies or a species is often quite arbitrary; the grouping of species into genera again is sometimes clear and objective, but often also arbitrary. How, in the face of such a graded and indefinite series, can we decide that subspecies are produced by one kind of evolution and species by another; or subspecies and species by one kind, genera by the other?

The geological time scale may be called in as an apology. It is true that the actual process of evolution, from the first reproducing protoplasm to the complex orchid or primate, has taken an immensity of time. But it is a sterile doctrine to believe that because of this the process is outside of the scope of laboratory study or analysis. The formation of mountain chains, the erosion of river valleys, also requires an immensity of time, but we can piece the process together from the fragments that we are able to observe and measure. Surely presently we shall also be able to do this for the process of evolution.

Two aspects of reproduction—the sexual process and the particulate basis of inheritance—are important in fitting together a picture of evolution, just as they are important in understanding the interrelations among individuals and populations of organisms, which is our immediate objective in natural history. The development of the individual, which forms the subject of the next chapter, is essentially an extension of the phenomenon of reproduction, since the whole process involves the projection of the final adult form of one generation to the final adult form of the next generation.

CHAPTER VI

The Development of the Individual

EVERY child is fascinated by the problem of the caterpillar and the butterfly. But most of us, as we grow up, forget these things and become absorbed in matters of consequence. Scientists form one group of individuals in whom childish traits persist: for the adult scientist still wonders about the problem of the caterpillar and the butterfly. He has never grown out of that exasperating period of childhood characterized by the eternal, "why, mummy?"

Perhaps creative genius is simply the persistence of childishness into adulthood in certain individuals: the persistence of curiosity to make scientists and philosophers; of wonder to make poets and painters; of jealousy and dominance to make leaders, statesmen and generals. Of imagination, of dream stuff, in all of these various categories: though it may be wiser not to pursue the childish analogy too far.

To get back to the caterpillar and the butterfly, it represents a case of metamorphosis, of an extreme type of solution of the problems of growth and development. Growth in organisms is never simply increase in size. The growth of crystals in a saturated solution, the growth of the delta of a river, the growth of a pile of garbage, can be thought of as mere quantitative accretion of materials.

But the growth of an organism involves qualitative changes, changes in proportions and in functions, so that the word has shifted its meaning. It might be better to speak of development with organisms.

Growth and development look quite different, of course, depending on the level of thought at which the phenomena are examined. It is possible, for instance, to study these things at the molecular level, which would be the point of view of the biochemist; or at the level of cells, which I suppose would be the point of view of the general physiologist; or at the level of individuals, with which many biological sciences are preoccupied. Individuals in turn are aggregated into populations and communities, which also show growth.

THE CONCEPT OF THE INDIVIDUAL

Growth, development and metamorphosis may gain perspective if we stop first to examine this question of the individual in more detail.

The universe of organic nature is made up of individual organisms. In a few forms with colonial habits, it may be difficult to determine precisely the limits of the individual—in the tumbling, spherical colonies of the protozoan Volvox, it is perhaps debatable whether the colony or the cell represents the individual—but such instances are rare. Generally, the individual is a concrete, objective unit, like the cell and the molecule. In some cases, the bacteria and protozoa, for instance, the individual and the cell coincide. In some of the viruses, like the virus of tobacco mosaic that Stanley crystallized, it seems that the individual, the cell and the molecule coincide.

At the level of biological thought which concerns us in this book, we are interested in the discovery, description and explanation of the various relationships that occur among individual organisms. Reproduction covers one such group of relationships, especially the sexual relationship and parent-offspring relationships, which bind individuals together into populations, or species. From another point of view, individuals of many different kinds may be tied together through food chains, shelter, proximity and common environmental problems, into biotic communities. With populations and communities, we reach concepts that are more subjective: hazy, indefinite, hard to define.

In sexually reproducing organisms, the new individual starts with the fertilized zygote and stops with death. With asexual reproduction, the span of the individual is not so clear, and it may be argued that the protoplasm of such an organism is immortal. But the germ stuff has continuity, immortality, with all organisms. The individual in asexual reproduction may perhaps most conveniently be considered to start with the fission that gives it birth, and to stop with death or with the fission whereby it loses its individuality.

If our interpretation be accepted, that the individual is the organism between each fission or other reproductive process, its life span may be very short indeed. Under favorable conditions, some bacteria may divide every 20 or 30 minutes, so that this time would represent the average life of the individual. A particular bacterium may, of course, go into the special state called a spore, and its life in that condition may continue indefinitely; but it is a state in which life processes are suspended, so that time has lost all biological meaning.

The life cycles of microorganisms sometimes follow very complicated patterns, but the complications involve, by our definition, a series of individuals, rather than form changes of a given individual. This would be the case, for instance, with the malarial parasite, Plasmodium, discussed in the previous chapter. In the human blood stream, the life of each individual parasite would be from fission to fission, and dissemination from one man to another through the mosquito cycle would involve a sequence of individuals: the asexual cell producing a gametocyte; two gametes fusing to form a zygote (here one could debate as to whether the zygote represented a continuation of the individuality of either gamete precursor); this subdividing into a large number of sporozoites; these in turn giving rise to new cycles of asexual individuals once they become injected into a new vertebrate host. The individuals in such a series would differ not only in form and function, but also in life span. The life of an asexual individual (it would depend on the species of Plasmodium) might be 24 hours; of a sporozoite, lingering in the salivary gland of a mosquito, several weeks.

This calls to mind the difference in significance of astronomical time for different types of organisms. The curve of growth of a population of bacteria, in which fission is occurring at an average interval of thirty minutes, may be very similar to the curve of growth of a human population except for the shift in time scale required to accommodate an organism with a generation every thirty years. Thirty minutes with such a bacterium corresponds to thirty years with man. Both populations are composed of individuals, and the aggregates of individuals appear to follow similar laws. But the scale of time has completely changed: a bacterial year is a human minute. This concept is recognized in the common observation that a month in the life of a dog corresponds to a year in the life of a man.

THE DEVELOPMENT OF MANY-CELLED INDIVIDUALS

The great difference between growth in single-celled and in multicelled organisms appears when we examine the process at the cellular level. Growth in a single-celled organism, reproducing by fission, seems to be similar to inorganic growth, to involve mere increase in size. Each new individual swims off, complete and functioning, faced only with the problem of getting to be big enough to split again.

In multicelled organisms, however, the complex adult individual must be built up from the single cell of the zygote by a process of repeated cell divisions, the various cells becoming organized into tissues with different functions, and the tissues coordinated into organ systems. Almost universally, the individual is protected during the first part of this process of development, and not launched on an independent existence until some of the essential organ systems have been built up to the point where they can function. The individual, during this protected period, is called an embryo.

In very simple organisms (like some coelenterates) the individual may be independent for its whole life. The gametes are released into the water and the zygote must fend for itself from the moment of fertilization. Even in coelenterates, though, there is apt to be a period of protection, the very early zygote living, for instance, in special "brood chambers" of the jellyfish.

When both kinds of gametes are released into the water, the female or macrogamete generally, even with very simple organisms, carries with it a certain amount of stored food material, of yolk, that can be used as a source of energy during the first cell divisions until the new individual has grown enough to develop methods of food collection for itself. The food adaptations of this very young organism may, because of the exigencies of the simple organization, be very different from those of the complex full-grown organism. Thus we get eggs, embryos, larval forms, and all of the complexities of growth and development.

THE EMBRYO

The earliest stages of such developmental processes are called embryos. They are the subject of a special science, embryology, to which every pre-medical student and every university biologist is exposed. The embryologists are interested primarily in explaining the genesis of the different organ systems, and the study is closely bound up with that of anatomy, histology and physiology. From the point of view of natural history we are more concerned with provision made by the parent for protecting the developing embryo, and with post-embryonic development, when the individual takes an active part in the biotic community.

I have been trying to think of some neat and general definition that would serve for "embryo", but without much success. With mammals the embryonic period ends with birth: but the stage of development at which the individual is born varies with different kinds of mammals. The extremes, perhaps, are the helpless, worm-like new-born opossums and the almost self-sufficient new-born deer or horse. With animals that lay eggs provided with yolk material, the embryonic period ends with the hatching of the egg. With seed plants, the embryonic plant is enclosed in the seed, and post-embryonic development would start with the sprouting of the seed, though the new plant continues for some time to use the stored food materials of the seed. In other major plant groups, the concept of embryo seems to have no meaning at all.

It cannot be said that the embryonic period ends when the individual ceases to be dependent on its parent or on material stored by the parent: look at the long post-natal period when the mammalian young are completely helpless by themselves, or think of the newly hatched birds that can do little more than open their mouths so mother can stuff down the food. Yet new-born mammals or newly hatched birds are not considered to be embryos.

At the other extreme, it is customary to speak of the very early free-living stages of coelenterates and starfish as "embryos", though "larvae" would be a better term, even for the very earliest stages, as soon as they cease to be dependent on yolk materials.

Perhaps the major landmark in development is not birth or hatching, but the period at which the new individual ceases to be dependent on the parent. In the vast majority of animals, this coincides with hatching from the egg and thus with the end of the embryonic period. The parent has provided an egg stored with yolk and has either turned loose an immense number of these eggs on the chance that some small proportion of them will lodge in some favorable situation and survive until hatching; or the parent has produced less eggs, but placed each carefully in some appropriate place where there is a minimum risk from enemies, and a probability of food for the new individual when it hatches. In either case, the new individual no longer depends on the parent once it has used up the yolk supply and hatched.

The larval starfish, the caterpillar, the young snake, has to shift for itself from the moment that the last of the parental yolk has been used up. But the young bird still has to be fed by its parents, and the young mammal depends on its mother's milk for a long time after birth. Post-embryonic dependence on parents is not confined to vertebrates. Think of the wasp larva that hatches in a cell carefully provided with paralyzed spiders by its parent, or the bee larva in a cell filled with the proper amount of pollen and honey. In such cases, dependence on the parent is complete until the new individual has become adult.

THE ADULT

Adult seems easy: my first impulse is to define it as the stage at which the reproductive mechanism begins to function. This, however, won't work, since quite a few organisms reproduce before they have reached what is called the "adult" stage. We cannot say, either, that adult is the period at which growth ceases, because there are many organisms in which growth ceases only with death. Best, perhaps, would be to say that the adult is the final form in the series of changes that may take place during development; and to observe that the assumption of this adult form generally more or less coincides with the beginning of reproductive functions; and that in a great many types of organisms, the adult form and the beginning of reproduction are closely related to the attainment of maximum size by the individual.

LANDMARKS IN DEVELOPMENT

We may not have achieved any very elegant definitions by this process, but we have marked a number of crucial stages in the history of individual development: hatching or birth (or sprouting); breaking of direct dependence on parent; beginning of reproductive function; attainment of definite, final form; attainment of maximum size marked by cessation of growth. It then becomes interesting to see how these various stages are related in different types of organisms.

During the embryonic period, food for the developing organism is supplied either in the form of yolk or (in mammals) from the parental blood stream. The problem of food gathering thus does not exist, and the problems of digestion and food absorption take a very special form. Some new mechanism for food gathering and food absorption must then be ready to function at the time of hatching or birth.

The organism also has no problem of self-defense during the embryonic period. While inside of the body of the parent, the problems of protection devolve completely on the parent. If the embryo is enclosed in an egg laid by the parent, it may be protected by an egg shell and may also be actively defended by the continuing presence of the parent; or it may be one of an immense number of eggs freed on the statistical certainty that a sufficient percentage, however small, will survive the chance hazards of the environment. In either case, the embryonic organism itself has no defense; and again some mechanism must be ready to function at hatching or birth.

In plants, the tie with the parent is broken early in the development of the individual, with spore formation or with the dispersal of seeds. Defense is almost always a matter of sheer numbers, of chance escape, so that the quantity of spores or of seeds must be immense. Since nutrition in plants depends on the absorption of dissolved materials, it is not apt to differ greatly in kind during development. The special plant phenomenon of the alternation of sexual and asexual generations has already been described.

In most animal types, the breaking of direct dependence on the parent coincides with birth or hatching, so that the mechanisms for food gathering and digestion and for self-defense must be completely developed and independent of parental control. In two phyla, the arthropods and the vertebrates, numerous types of parental dependence have been developed that may continue long into the post-embryonic period.

The newly hatched organism may be very unlike the adult: it may have a totally different form, and be adapted to a totally different environment. Development to the adult form may be through a gradual series of changes, of the sort that we normally think of as organic growth; or the changes may be abrupt, like those of the caterpillar and the butterfly, constituting a metamorphosis.

Metamorphosis is commonly defined as a conspicuous change in form and mode of life occurring in a comparatively short time and not directly associated with an increase in size. Such abrupt changes during development are found in several animal phyla. They represent solutions of the problem of bridging the gap between the relatively small and simple post-embryonic organism and the complexly adapted adult. A sort of metamorphosis occurs in the protozoa and coelenterates, described in earlier chapters, but this by our definitions involves an alternation of individuals or a succession of individuals, and thus is not a part of the problem of individual development.

THE METAMORPHOSIS OF INSECTS

The classic examples of metamorphosis are found in the insects and the amphibia; and in both of these cases the change is associated with a shift in the environment to which the organism is adapted. Growth in all insects is divided into sharply defined periods because of the necessity of shedding the hard external skeleton for increase in size. The skin of an insect is made of hard stuff, chemically somewhat similar to the material of our finger nails, and serves both for support (skeleton) and for protection (skin). The material is not composed of cells, but is secreted by cells; and once formed, increase in size (growth) is limited to stretching in flexible areas (joints). The organism grows until it has stretched its skin to the maximum possible extent, then a new and larger skin is secreted under the old one; the old one is split, shed and discarded; and the organism is then ready to grow until it again reaches the limits of stretch of its new skeleton-skin, when the whole process must be repeated.

Changes in form in an insect are thus necessarily associated with the periodic moultings when the formation of a new skin gives an opportunity for the formation of a different armor. The successive differences are not striking in many insects: a grasshopper, after each moult, looks more or less as it did before except for changes in proportions and except for the development of functional wings and of reproductive organs along with the last skin. In many orders of insects, however, the evolution of the young post-embryonic organism and of the adult have followed ever more diverging courses until we get the striking differences shown by the caterpillar and the butterfly, or by the maggot and the fly.

This "complete metamorphosis" probably originated among insects that had acquired adaptations to fresh water. The fresh-water environment offers all sorts of possibilities for organisms—fresh water is plentiful, it contains much food material, it is less subject to temperature changes than air, and organisms in the water are not faced with the problem of how to keep their body fluids from being lost by evaporation—factors that will be discussed in more detail in later chapters. Its great disadvantage is transience: a puddle may dry up in a few days, a marsh in a few years, a lake in a few thousand years. Fresh water thus does not offer continuity, and it has not been the scene of long-term evolution of major organic types. The animals adapted to fresh water represent a very great variety of types that have undergone their basic evolution in the sea or on land, and that have subsequently found means of adapting themselves to fresh water.

Insects apparently discovered the advantages of fresh water early in their geological history. The adult, winged insect has particular advantages for searching out different ponds, puddles and streams; and these environments offer food and protection for the young, without competition from a well adapted pre-existing fauna.

The first adaptations in insects for life in fresh water were probably of a sort that enabled them to invade the water temporarily. Many aquatic insects today get oxygen from air in a bubble that they carry down with them in diving, or from air kept in a film around the body by hairs, and these get along about equally well in or out of the water. Many, however, have developed gills of one sort or another for getting oxygen directly from the water, or elaborate apparatuses like the breathing tubes of mosquito larvae for getting oxygen from the surface air—specializations that cripple the insect when it is out of the water, so that it has become confined to the fresh-water habitat.

Adult insects have remained primarily terrestrial animals. They have solved the problem of flight, which gives them immense advantages for dispersal, for seeking out appropriate growing places for their young. It is easy to imagine, under these circumstances, how the young and the adult of the same kind of insect could become, in the course of evolutionary history, increasingly different. The aquatic habitat offered great possibilities to the immature insect, the terrestrial habitat great possibilities to the adult, but the adaptations for successful life in the water and for life in the air are very different. Once this divergence in mode of living was started, the young would become subject to different environmental pressures from those affecting the adult, which would tend to set up divergent evolutionary trends.

The young and the adult would thus become increasingly different, and the difference would become increasingly difficult to bridge in the single skin change with which the animal left the water and took to the land. In fact, among living insects, a whole series of increasing complexity can be made out, from species that make the change with a single moult of their skin, to those that require a long rest period during which the whole structure of the body is rebuilt.

This rest period is called the "pupa." It is a unique invention of the insects, and its intercalation in the life history has made possible the extreme of divergence between the young and the adult represented by the caterpillar and the butterfly. Through this mechanism the young animal, the larva, becomes specialized for the acquisition of food, which is needed in large amounts for growth; it also generally stores food materials as fat, to be used up during adult life. The adult, then, may become more and more specialized for the functions of reproduction and dispersal. In many insects, the adult takes no food at all, and in other cases it takes but small amounts, living chiefly on fat carried over from the larval stage.

The larva, with such insects, is a very different sort of organism from the adult, a sort of perambulating embryo carrying the adult characters as undeveloped bits of tissue scattered through its body. When the larva has attained full size, it goes into the pupal state, in which all of the special structures of the larval are broken down, and the development of the adult insect starts in anew from the buds of embryonic tissue. The pupa, like the egg, is defenseless, and various mechanisms have been developed for protection during this period. The larva may spin a cocoon for the pupa, it may develop a very hard armored skin, it may burrow into the ground or find some other situation in which it will be inconspicuous and protected.

THE METAMORPHOSIS OF ECHINODERMS AND AMPHIBIA

The echinoderms undergo a metamorphosis as striking as that of the insects. The larval sea urchin or starfish is a free-swimming, transparent, bilaterally symmetrical organism. The change from this larval form to the radially symmetrical, hard-skinned, slow-moving adult form may be abrupt: in the case of some sea urchins, taking place in a half an hour. When ready to metamorphose, the larva sinks to the bottom and anchors itself with "tube-feet" which have been formed under the larval skin; the ciliated arms of the larva seem to melt away; the body fluid is expelled, and the body becomes compressed into a flat disc which then crawls away as a young sea urchin.

The physiology of metamorphosis in the amphibia—the change from tadpole to frog—has been studied in great detail. The process in amphibia is controlled by hormones secreted by the thyroid gland, and if the thyroid is removed, the larva will never metamorphose; or, contrariwise, metamorphosis can be hastened by feeding the larvae with thyroid extract from any kind of vertebrate. Different species of amphibia differ greatly in the speed with which metamorphosis occurs, but the process in all cases is slow as compared with the metamorphosis in the invertebrates.

PARENTAL CARE AND THE LIMITS OF GROWTH

Continuing dependence of the post-embryonic young on the parent occurs sporadically in many groups of organisms, but it is especially characteristic of the order Hymenoptera (ants, bees, wasps) among the insects, and of birds and mammals among the vertebrates. In the bees, wasps and ants, the young are completely dependent on the parent until they in turn are adult, and this is almost true also in man. The dependence seems, in both cases, to be associated with the development of the social habit. The social habit, at least in the Hymenoptera, appears to be a result of the parental problem of taking care of the young, rather than the reverse, as is shown by the various intermediate stages among wasps. The adult may provide all of the food for the larva, though continuing itself to lead a solitary existence, and the line of evolution is from this to the very complex societies of the most highly developed forms.

In birds and mammals, the length of time that the young is dependent on the parent varies greatly in different species, and the shift to an independent existence is almost always gradual.

Growth in many animals and plants has no definite limit: the organism continues growing as long as it lives; though with each structural type of organism there is a vaguely defined limit to possible size. The rate of growth in such organisms differs greatly in different species, as does the average length of life, so that some kinds are always small, while others become giants. The reproductive system may become functional long before the average maximum size is reached: this is obvious in trees, reptiles, fish, to take some instances at random. In the case of organisms with a fixed adult size, the reproductive system may start to function either before or after this size is attained. In mammals, for instance, the animal is capable of reproducing before growth has stopped; in many insects, however, the reproductive system does not become functional until some time after the fixed adult stage has been reached.

THE RECAPITULATION THEORY OF HAECKEL

The very early stages of development are similar in the most diverse kinds of animals. The zygote, or fertilized egg, divides into two cells, then four cells, then eight, and so on, until there is a small hollow ball of several hundred cells, called the blastula by embryologists. As cell multiplication continues, the ball becomes indented at one point, the cell layer pushing in until eventually there are two layers of cells surrounding an open cavity. This stage is called the gastrula. Development after the gastrula diverges more widely in different phyla, the differences depending chiefly on the sort of provision made for feeding the embryo, but in general the similarities between embryos in the early stages are more striking than the differences. The embryos of a chicken and a man, for instance, do not differ markedly until the organisms have reached a considerable degree of complexity.

Embryos are similar to each other. But also the embryos of complex animals may resemble the adults of simpler animals. This can be carried all the way back, the zygote being compared to a protozoan, the blastula to a colonial protozoan, the gastrula to a coelenterate. The embryos of birds and mammals go through a stage with structures called "gill slits," and these have been taken to represent traces of an ancestral fish stage.

These resemblances formed the basis of a famous generalization developed by Ernst Haeckel, in books published in 1874 and 1875, and known as Haeckel's recapitulation theory. It is usually expressed in the elegant words of the statement that "ontogeny recapitulates phylogeny." Ontogeny means the sequence of development of an individual from zygote to adult. The caterpillar is thus an ontogenetic stage in the development of the butterfly. Phylogeny means the evolutionary history of a kind of organism. The primitive mammals of the Cretaceous represent a phylogenetic stage in the development of modern mammal types.

In other words, Haeckel said that an animal, in the course of its development, goes through a sort of synopsis of the evolutionary history of the species, the various stages representing various ancestral types.

This theory at least had the virtue of stimulating a great deal of work in embryology. The paleontological record was hopelessly incomplete, but here, in the development of living organisms, might be the clue to the path of evolution in the various groups.

The theory has, however, not been sustained. We now realize

that the essential point is not that the young of complex animals resemble the adults of simpler animals, but that rather the very young stages of all animals are similar. The early stages of complex animals do not correspond to the adult stage of simpler types that were ancestral; they correspond to the equivalent early stages of simpler animals. The various animal types diverge more and more widely during the course of growth and development, all being similar at first, but differing more and more as the special characters of the different phyla, classes and orders appear.

This might be expected, since the protected early stages of animals would be less subject to environmental pressures than the later, free-living stages. The environment within the egg shell or within the parental ovary is about the same in all sorts of different animals, and the problems of cell multiplication and growth remain similar until the organism reaches the stage where the special adaptations of its particular group must begin development.

The similarity of the embryonic stages of very different kinds of animals was first observed by Karl von Baer, who lived from 1792 to 1876, and thus preceded Haeckel by many years. Von Baer carried out a great deal of detailed study of animal development, discovering among other things the egg of mammals. He formulated his observations on the similarity of early stages of animals, and on the sequence of development whereby general characters appear first in ontogeny, special characters later, in a series of statements which have come to be known as "von Baer's laws." These form a more practical guide for developmental studies than does the specious "recapitulation theory" of Haeckel.

NEOTENY: THE CHILDISHNESS OF MAN

Actually there is considerable evidence that a process more or less the reverse of that postulated by Haeckel may have been important in evolution: a process called "neoteny." This word is used to cover those cases in which the adult form of an animal resembles the young form of its presumed ancestors; or, in other words, cases in which characteristics of the young of the ancestor are retained by the adult of the descendant. It may be interesting to explore this theory a little, since neoteny bears directly on the problem of human evolution. We seem to be a sort of ape that never grows out of its childish traits.

Numerous cases are known in which the reproductive organs may become functional during the larval stage. The famous Mexican salamander, the axolotl, forms the classical case: this was long thought to be a special kind of animal until it was discovered that under certain conditions (feeding thyroid, for instance) it grows up into an ordinary salamander of the genus Amblystoma. Certain other newts have apparently completely lost the ability to grow into the adult form, even under experimental conditions. Again, certain insects have been known to reproduce as larvae; and some, such as the female glowworm, appear to be larvae that have lost the ability to grow up.

The significance of this in evolution is speculative, but none the less interesting. The adult echinoderm—the starfish or sea urchin—is surely not ancestral to any of the more complex animal types, but the larval echinoderm shows striking resemblances to some of the simple chordates, the presumed precursors of the vertebrates. Again, a larval myriopod, soon after hatching, looks much like the adult form of simple insects.

The theory of the relation of neoteny to human evolution was most extensively developed by L. Bolk, in a book published in 1926. Bolk showed that many of the adult features of man resemble most closely embryonic structures in apes: the relatively high brain weight, the angle which the head makes with the trunk, the retarded closure of the sutures between the bones of the skull, the dentition, the flatness of the face, the hairlessness of the body, the light color of the skin, and so forth.

The sutures between the bones of the skull in man do not close until he is about thirty years old. In other mammals, including the apes, these sutures close much sooner after birth, and when that has happened, the skull can no longer increase in size. The brain of the new-born chimpanzee resembles that of adult man in various respects more than does that of the adult chimpanzee.

The hair business is particularly interesting. This is summarized by Bolk (quoting from de Beer) as follows:

1. The monkey is born with a complete covering of hair.

2. The gibbon is born with the head and back covered with hair, and the other regions are covered later.

3. The gorilla is born with the head covered with hair, and the other regions are partially covered later.

4. Man is born with the head covered with hair, and the other regions are scarcely covered at all later.

The very fine hair which covers the unborn infant is also present in the unborn apes, and is not completely shed in man until after birth.

Thus a whole series of human characters can be explained by a retardation of the rate of development in man as compared with his primate cousins. The development of the reproductive organs is also retarded in man, but the human characters are explicable if we assume that the reproductive organs become functional relatively late, but while the other body characters, as compared with the presumed ancestral type, are still juvenile.

The theory of neoteny provides, at least, a fruitful field for speculation. Most of us find young mammals much more fun than the adults, since the adults necessarily become dull and serious with their numerous responsibilities and preoccupation with the problems of existence. Perhaps it is because of this that I find a particular pleasure in thinking of man as a kind of mammal that has lost the ability to grow up.

CHAPTER VII

The Environment

NATURAL history is often considered to be more or less equivalent to the special science of ecology; and ecology, in turn, is generally defined as "the study of organisms in relation to their environment." Environmental relations, then, form the essential core of natural history, which we have been approaching in a leisurely and roundabout way.

These first chapters have been concerned mostly with background material. An explanation of the naming system of organisms seemed advisable at the start. This led to a rough sketch of the major divisions of the classification of organisms, the filing system used for handling natural history observations. A logical next step seemed to be a review of the history of the development of these major divisions, as we know it from the fossil record. The continuity of the history of organisms leads naturally to a consideration of the phenomena of reproduction, and reproduction to growth and development, the subject of the previous chapter.

From this point, the consideration of growth and development, we might proceed to the study of the problems of adaptation and evolution, but before taking that step, we need to examine the characteristics of the environment itself, so that we can better understand the ways in which environmental factors influence adaptations. This involves not only the physical environment, but the biological environment: the relations of organisms among themselves, in communities, in mutualistic associations (symbiosis) and in antagonistic associations (parasitism and predatism).

THE DEFINITION OF ENVIRONMENT

The environment is normally defined as the sum of the external forces acting on an organism. It is thus a very broad concept and, as Charles Elton has pointed out, it is not always easy to decide where the organism ends and the environment begins. The immediate environment of the embryonic organism, for instance, is the womb of the parent, the materials arranged by the parent in the egg, or the material and arrangement of the seed. The details of this environment may be tremendously important in controlling the development of the embryo. The experimental embryologists have shown this by tampering with the embryonic environment and thus upsetting the normal sequence of development.

Even in the independent adult the line between the organism and the environment, between internal forces and external forces, is not always easy to draw. The animal or plant and the environment, in fact, form an interdependent system, each modifying the other. Thus the growing rootlet of a plant controls, to a considerable extent, the chemical and physical nature of the soil in the immediate vicinity of the rootlet. Or, to take an extreme case, consider the effect of human civilization on the appearance of the countryside. New York City is an environment created by man; but this environment in turn affects man just as surely as the natural environment of the arctic affects the Eskimo.

Food is a basic component of the environment. But where, in the digestive process, does food cease to be part of the environment and start to become part of the organism? In the mouth, in the stomach, in the blood stream? The human intestine forms,

surely, the immediate environment of an intestinal worm; but is not
the worm also a part of the environment of the man?

This is quibbling, except that it illustrates the difficulty, perhaps the danger, or even the futility, of logical definition. The
organism—its form, history, activities, potentialities—is the product of forces which, on analysis, derive from two different sources:
one internal, stemming finally from the germ plasm; the other
external, representing the environment. The difference, at the level
of natural history, is clear and profound. At another level of
thought, this difference might disappear, since surely the germ
plasm is also in some way the product of forces external to itself.
But with this line of thought, we enter into the area of the problem
of the explanation of biological phenomena in physico-chemical
terms.

CLIMATE

It is convenient to think of the environment as including biological factors and physico-chemical factors. The principal physical
environmental factors, in the case of terrestrial organisms, are covered by the expression "climate."

Climate could be defined as the sum of the variables of the
atmosphere: rainfall, temperature, wind, sunshine or cloudiness,
solar radiation, humidity. The influence of this atmospheric climate
on aquatic organisms is indirect. Water is subject to similar variables, but we do not ordinarily think of them as making up a special
aquatic climate.

The word climate comes from a Greek root for "incline,"
referring to the slope of the earth toward the pole. A change in
clima then meant a change in latitude, which was gradually seen
to mean a change in atmospheric conditions and length of day, so
that the word acquired its present meaning. "Weather," in current
usage, denotes a particular climatic event; and the climate of a

particular place could thus be thought of as the average of its weather.

It is useful, in biological work, to think of climate in this ordinary sense as the "geographical climate," thus remembering that there are other levels at which atmospheric conditions need to be studied. The geographical climate is described by the measurements taken at standard meteorological stations: the rain is measured by a particular model of apparatus placed so that the rain reaching it is never diminished by obstructions; the temperature is measured in a particular kind of a box at a fixed height above ground; the moisture content of the air is measured at fixed hours of the day, and so forth.

By standardizing measurements in this way, atmospheric conditions at a given place can be compared at different seasons, and by averaging conditions at a given place, geographical comparisons can be made—between Miami and Jacksonville, Maine and Texas, Massachusetts and Germany, or North America and Africa.

If we stop to think, we realize that very few organisms live under the conditions measured by these standard meteorological stations. Temperature and moisture conditions in a forest are very different from those in the open field where the measurements are taken; they may be different in different kinds of forest, or in different parts of the same forest. The relevancy of these standard measurements is even less if we think of the atmospheric conditions surrounding a particular individual organism: a grasshopper on the under side of a leaf, a mushroom growing on a rotten log, a gopher sleeping in its burrow.

For most biological purposes, the geographical climate is useful chiefly as a base line with which these different special climates can be compared, and as a sort of index to the general types of conditions that are liable to be found in a given region. It is useful for the naturalist to think of climate at two other levels. Atmospheric conditions in a particular landscape type, such as a forest or a savannah or a mountain ravine, might be called the ecological climate. At the final level, that of atmospheric conditions as they affect a particular individual organism, it is convenient to speak of the microclimate.

Climates, at whatever level, involve variations in four main sets of factors: temperature, water, light (radiation) and air movement. The gaseous composition of the air, which might conceivably be another factor, is so constant that it is left out of consideration in discussions of climate.

TEMPERATURE

Temperature is of great importance in all life processes. Birds and mammals, by achieving considerable control over body temperature, have made themselves independent of environmental temperature to a greater extent than other groups of organisms; but even with birds and mammals, the range of temperature that can be tolerated by a given species is definitely limited. If the body temperature of the organism drops below the freezing point of water (0° Centigrade), all normal life processes stop, and with most organisms, these processes are suspended before this temperature is reached, at seven or ten degrees Centigrade. Only a few organisms, with very special adaptations, can survive when their body temperatures exceed 45° to 50° Centigrade (more or less 120° Fahrenheit). When one thinks of the range of temperatures known to exist in the universe, these are very narrow limits indeed.

WATER

The importance of temperature to life depends on the physical properties of that peculiar chemical, water. Life is an aqueous solution, and a good deal of biology is concerned with the behavior of the solvent, water. I wish someone with an adequate background of biology, physics and chemistry would write a book about water.

It is a theme that could be used to unify all of the sciences that are concerned with things and events at the surface of the earth.

There are many reasons for supposing that life arose in water, specifically in the waters of the pre-Cambrian seas which had perhaps a third or so of the salt content of sea water today. The ratios of inorganic salts in protoplasm seem to have been fixed by the salinity of these pre-Cambrian seas and remain remarkably similar today in the most diverse sorts of organisms.

Much of evolution can be interpreted in terms of modifications of organisms to enable them to live in environments differing from these pre-Cambrian seas: slow and slight modifications for life in the increasingly saline seas of later periods, more drastic modifications for life in fresh water, and most drastic of all, modifications for life on land. The majority of animal phyla are still purely aquatic, and most animals and plants that have invaded the terrestrial environment are dependent on obviously damp situations.

This is perhaps clearest in the mechanisms of sexual reproduction. The gametes of mosses and ferns must still meet by swimming, even though the swimming performance may be confined within a drop of dew. The seed plants, to avoid this need for free water, have developed the curious gametophyte phase of pollen which is brought near the macrogametes by elaborate behavioral adaptations, such as insect pollination, so that fertilization can be achieved through the mechanism of the pollen tube. Animals, particularly of the two great terrestrial phyla of arthropods and vertebrates, have developed equally fantastic behavior patterns culminating in intromission, so that the gametes, still completely dependent on the liquid medium, can reach each other.

Terrestrial life, then, depends on water; and a great deal of the behavior and physiology of terrestrial organisms is concerned with how to get water, how to retain water, how to economize in the use of water. In this they are tied up with the gas-liquid-solid cycle of water which forms such an important part of climate.

Climatic measurements of water take various forms, but especially involve precipitation (rainfall, snowfall), relative humidity and evaporation. Precipitation needs little comment here. There are a few areas of the world where rain is an almost unheard of phenomenon (parts of the Sahara, of the Peruvian coast), and there are wide areas where rain is so scarce that what little life exists depends on very special modifications for desert conditions. The seasonal distribution of rain may be more important than the annual total in determining the character of the regional biota— this is especially marked in those parts of the tropics where there is an alternation of wet and dry seasons. Adaptations for survival through the dry periods are often comparable to adaptations for survival during the unfavorable cold period of temperate latitudes.

The highest known annual rainfall is 451 inches, the average for Kauai, one of the Hawaiian islands. Tropical rain forest, which from its luxuriance and variety must represent the most favorable of terrestrial environments for life, requires a rainfall of at least 150 inches (more or less 4 meters) fairly evenly distributed through the year. Most parts of the world receive an annual rainfall somewhere between 30 and 80 inches.

HUMIDITY

The biologist is as much interested in relative humidity and evaporation as he is in precipitation. The amount of water vapor that a given volume of air can hold in suspension depends, of course, on the temperature of the air. For this reason, it is customary to measure the water content of the air in terms of relative humidity: that is, to compare the amount of water vapor in the air with the amount that air at that temperature could hold if saturated. When we say the humidity is 90 per cent, we mean that the air has 90 per cent of the water that it could hold at its temperature at that moment. The actual amount of water in a cubic foot of air at 90 per cent relative humidity is thus very different on a hot day and on a cold day.

The rate of water evaporation would probably be one of the most useful of climate measurements for biological purposes, except that it is difficult to standardize, especially for regional comparisons. It is thus used more for studies of ecological climate than of geographical climate. Rate of evaporation depends on temperature, on air humidity, on air movement, and it sums up these various factors more or less in the same way that they affect organisms.

AIR CURRENTS

Air currents of various sorts are always measured by meteorologists, and they are important in analyzing the genesis of a given geographical climate, or in predicting climatic fluctuations. The influence of various sorts of air currents on organisms is often indirect and hard to evaluate.

In protected situations, such as the tropical forest, delicate leaf forms are developed that could not withstand exposure to wind. The form of trees on the coast or high on a mountain may be determined by the constant winds. Insects in exposed situations, such as oceanic islands, are apt either to lose the power of flight or to develop very strong flight, strong enough to surmount the usual wind. Many organisms depend on air currents for spread, as do spores and many kinds of seeds.

LIGHT

Light, very important for organisms, is not usually measured as a part of geographical climate except in general terms, such as hours of sunshine. Seasonal fluctuation in length of day—an astronomical component of geographical climate—is an important factor in determining the distribution and adaptations of organisms. The timing of seasonal biological events, such as the flowering of a particular plant, or the migration of a kind of bird, may depend on the length of day, instead of on the more obvious seasonal change in temperature. The geographical shift in length of day is tremendous, from the extreme of an annual succession of exactly even days on the equator to the midsummer total light or midwinter total darkness of the poles.

The prime source of energy for all life processes is of course solar radiation; and this is subject to seasonal and geographical fluctuation, and perhaps to long-term cycles. Radiation has, however, been relatively neglected in biological studies of climate.

TOPOGRAPHY

We defined climate as the sum of the variables of the atmosphere. The physical environment of terrestrial organisms would also include topography, since such organisms, though they breathe air, are also chained to the earth, the eagle as well as the oak tree. Climate and topography are intimately related, since mountain chains and coastal configuration may control precipitation, temperature, humidity and air currents; and since rain, wind and temperature, through erosion, may control topography.

The physical nature of the soil, of the substrate of the terrestrial biota, should also be included in this list of physical factors. Though here again it is difficult to separate cause and effect: the nature of the soil is an important factor in determining the type of vegetation, and hence also of animal life; but the nature of the soil is also, in large part, a result of the type of vegetation that has been growing upon it.

MARINE VERSUS FRESH-WATER ENVIRONMENTS

The physical factors of the aquatic environment are in part similar to those of the terrestrial environment, in part very different. In the first place, there are two major sorts of aquatic environment, the marine and the fresh water, which require very different adaptations on the part of their inhabitants.

The marine environment covers the major part of the surface of the globe. It has both geological and geographical continuity. It probably represents the environment in which life arose, and it has certainly been the scene of the greatest proliferation of life forms. All of the major animal phyla have representatives in the sea, and most of them had their origin in the sea. Only a few major groups, like the insects and the vertebrates, seem to have had their origin through adaptations to existence in fresh water or on land. On the other hand, the evolution of the major plant groups has more clearly been bound up with existence away from the sea, and the marine flora remains predominantly composed of relatively simple types, especially bacteria and algae.

It is interesting that while the sea has far more major types of organisms than any other environment, it falls far behind in numbers of kinds or species. The greatest proliferation of species has been on land, among the seed plants, the fungi, the insects and the vertebrates. On number of individual organisms, the sea would probably win again, because of the incredible numbers of microscopic organisms that live in its surface layers.

The fresh-water environment has neither geological nor geographical continuity. Our biggest and most impressive lakes are transient affairs in geological history, with two or three curious exceptions—Lake Baikal in Siberia, Lake Tanganyika in Africa and, to a lesser extent, Lake Ochrida in Europe. Baikal and Tanganyika have both great age and great depth, and they have developed extraordinary local faunas. Some of the major river systems of the world also have great geological age, and have developed correspondingly distinct faunas.

For the most part though, fresh water is transient in the geological sense, and much of it is transient in any sense. Lakes are formed by geological accidents and start at once on the succession of stages through which they gradually fill up with debris to form marshes or swamps or dry land. Rivers fluctuate tremendously in volume, as do many marshes and lakes, and many extensive areas of fresh water dry up seasonally, or irregularly.

The fresh waters of the earth are also discontinuous, so that each fresh-water system must be invaded independently from the neighboring sea or land through adaptations on the part of the marine or terrestrial biotas; or else the inhabitants of fresh water must have some means of dispersal through the intervening terrestrial or marine environments.

GENERAL CONSIDERATIONS ON THE AQUATIC ENVIRONMENT

The physical factors of the aquatic environment, whether marine or fresh water, are much more constant than those of the terrestrial environment. The high specific heat of water makes for slow temperature changes. Air temperatures may change with abrupt irregularity, and there are always daily and seasonal temperature cycles in the air. Corresponding changes in water are always less than those in the air, and the daily and seasonal cycles are pronounced only at the water surface, or in shallow water, or in small water accumulations. The temperature at different depths in the open sea is practically constant, and this is true of fresh water in the mid-tropics two or three feet below the surface.

Relative humidity, of such great importance to terrestrial organisms, does not exist in the aquatic environments; water loss through evaporation is not a problem. There is, however, another physical factor that is of great importance to aquatic organisms: the specific gravity of the medium.

The similarity of the specific gravity of water, especially sea water, and the specific gravity of protoplasm, means that aquatic organisms do not require massive skeletons to support their weight. There seems to be no physical limit to the size of a giant kelp, a squid or a whale, whereas land creatures are definitely bound by the problem of supporting and moving the accumulating weight of protoplasm that goes with giantism. A very large proportion of the aquatic biota—called the plankton—simply floats or drifts with the medium. Swimming is developed in aquatic organisms not for support, but for locomotion; whereas in the air, a constant output of energy is needed merely to maintain position, except in very tiny organisms such as spores or bacteria. Organisms on land are tied constantly to the surface of the earth; but life in the sea may be completely independent of the bottom.

The factor of specific gravity makes for the profound difference between the fresh and the saline environments. Aquatic organisms are not bothered by evaporation, but they are faced with the problem of osmosis. Any very clear explanation of osmosis would require a considerable digression; and I don't know whether such an explanation is needed or not. I suspect that osmosis is a word that most of us carry with us from high school science, without remembering at all clearly what it means. It refers to the diffusion of solutions through semi-permeable membranes, and it is an important factor in all physiological processes.

The body fluids of aquatic organisms are separated from the water in which the organisms live by various sorts of membranes—the skin, the digestive tract, the gills and so forth. The salts of the body fluids are in general more dilute than the salts in sea water, and more concentrated than the salts in fresh water. It is consequently important that the membranes that divide the solutions on the inside of the organism from those on the outside be of a sort that will not permit osmotic pressures unfavorable to the organism to develop. Since conditions in fresh water and salt water are so totally different when compared with the general condition of protoplasm, the two types of aquatic environment require different physiological adaptations on the part of their inhabitants.

Physical pressure of the medium is another important environmental factor in water. Air pressures vary with altitude, but the variation is slight as compared with the variation in water pressure with depth. Air pressures are important with terrestrial organisms only on very high mountains; but water pressure is an ever present factor in controlling the distribution of organisms in the sea.

I remarked that the gaseous composition of the air might conceivably be a factor in climate, except that it is quite constant because of the rapid diffusion of the various gases that make up air. In the water, however, dissolved gases diffuse much less rapidly, and the amount of dissolved oxygen or dissolved carbon dioxide in water is a very important variable in the aquatic environment.

THE CHEMICAL ENVIRONMENT

Dissolved gases might be considered a chemical rather than a physical factor of the environment. But as I remarked at the beginning of the chapter, any classification of environmental factors is pretty much an artificial convenience. In studying aquatic organisms the importance of chemical factors in the environment becomes particularly clear. Pure water does not exist outside of the laboratory, and it is very hard to make and to maintain even in the laboratory, since almost everything is liable to dissolve to some extent in water. Rain, by the time it has reached the ground, has picked up traces of many chemicals from dust in the air, as well as gases in solution from the atmosphere itself.

In studying terrestrial animals, one is apt to forget about the chemical environment, since its effect on such animals is indirect, acting through the biological factor of the food supply. Terrestrial plants, though, are as dependent on the chemical environment as are aquatic organisms, since they are in direct contact with chemical solutions in the soil.

The biological textbooks often list the chemical elements that seem to be necessary for protoplasm. Thirteen such elements are commonly found in considerable quantity in organisms: hydrogen, chlorine, oxygen, sulphur, nitrogen, phosphorus, carbon, silicon, sodium, potassium, magnesium, calcium and iron. Many other elements are found in traces in all sorts of organisms, and are necessary in many cases for the survival of the organism, but the quantities involved are so small that study and analysis is difficult.

Almost all organisms get oxygen from the atmospheric gas either directly or in solution in water; and almost all organisms require considerable quantities of free oxygen and free water. The exceptions to this, such as organisms that have dispensed with oxygen, or found means of manufacturing water, are interesting but sporadic. Terrestrial animals get the other elements that they need along with their food supply, which stems ultimately from plants. Plants—or at least those that contain chlorophyl—are able to build up their protoplasm on a basis of carbon obtained from the carbon dioxide of the air, but for the rest of the elements, they are dependent on salts in solution in the soil or in free water. The relative concentration of the different elements and available salts in different types of soil and water then becomes very important in studies of plant ecology. Some plants may have the ability to pick up ions of a given element in very dilute solution, whereas others may require a considerable concentration. The distribution of a particular kind of plant may thus depend on the nature of the chemical environment.

This applies also to aquatic animals, since many of these require larger amounts of given elements than are obtained from their food, and thus depend on salts in solution in the medium.

The chemical content of the soil or water may also influence organisms in more indirect ways. Certain substances may be poisonous to particular kinds of organisms if present in large amounts. Or the organism may be affected by the acidity or alkalinity of the soil or water, which depends ultimately on the kind of salts present in solution.

I am treading here a very narrow path, trying to indicate the importance of the chemical environment without becoming involved, on the one hand, in complicated explanations that would be out of place in this book; and avoiding, on the other hand, oversimplification, or statements that would be erroneous or misleading. In considering the chemical environment and the chemical requirements of organisms, we become concerned with food. But food involves the relation of organisms to each other as well as to the chemical and physical environment, and hence is best considered as a part of the biological environment, as one of the relationships in a community of organisms.

THE BIOLOGICAL ENVIRONMENT

In studying the natural history of a particular organism, say a fish, it is convenient and relatively easy to sort the environmental factors into groups. In this case, temperature, water currents, water density, light, could be grouped as making up the physical environment. The various substances dissolved in the water would make up the chemical environment. The biological environment would include the organisms used as food, the predators and parasites that made up the list of enemies of the fish, the vegetation that provided hiding places and nesting places, and so forth.

Such groupings are less easy when one is trying to analyze the general concept of environment without reference to any particular kind of organism. In the case of the sort of things that might be listed as making up the biological environment, it seems to me better not even to attempt any formal classification of component factors.

The following three chapters are really a discussion of the biological environment. In the next chapter, I shall attempt to build up a general picture of the biotic community. The community might be taken as the total biological environment of any of its component members, but if it were described as such, emphasis would be different depending on which particular member of the community was under study.

The problem might be compared to that of describing a human community. The description of a city could be considered as the description of an urban environment. The city, or the environment, would look different, though, if one were studying transportation, the steel industry, housewives, carpenters, or the environmental relations of John Smith. For the present purpose, I don't want to describe the environmental relations of Smith, or of carpenters, or of transportation systems. I want to build up a picture of something that contains all of these things, a city. Or rather, its biological analogue, the community of organisms.

CHAPTER VIII

Biotic Communities

I HAVE been trying to think about an organism living alone, in isolation. It is not an easy condition to imagine, but perhaps the attempt will make a good start toward understanding the interdependence of organisms in communities.

Animals are out. It is impossible to imagine any kind of animal living alone, because all animals need fairly complex carbon compounds as food, and they can only get these by eating plants, or by eating animals that in their turn have eaten plants. So we are limited to the plant kingdom in our search.

The so-called "higher plants" are out too, if only on account of nitrogen. An oak tree or a dandelion can build starches from water and atmospheric carbon dioxide, but such plants are dependent also on various chemical elements from the soil. Among these necessary elements is nitrogen. It is one of the commonest elements, making up as a gas the major bulk of the air; but neither the dandelion nor the oak can use nitrogen in its simple form; it must be combined into some soluble salt such as a nitrate.

The inert nitrogen gas of the air may be "fixed" as oxides by lightning flashes, and washed down to the soil as nitrous acids by rain; but this still is not in a form that can be used by the dandelion or oak. It must be oxidized into the nitrate form by soil bacteria. We don't know of any way these "higher" plants can get their nitrogen supply without the intervention somewhere of microorganisms. Plants of the bean family, as every farmer knows, can be used to build up the soil nitrates, but this is because the beans act in cooperation (symbiosis) with certain special bacteria: the beans, isolated from the bacteria, would be helpless.

But what about a bacterium living in isolation? Bacteria do not have chlorophyl, and thus cannot build up the carbon compounds necessary for protoplasm with solar energy, as the green plants do; most bacteria are thus dependent on green plants for their carbon compounds. A few kinds, though, have exploited completely different sources of energy, and some of these seem to be completely "autotrophic," able to live on purely inorganic materials of the sort that might be presumed to exist in the complete absence of other life. Such are the sulphur bacteria, which obtain energy from the oxidization of sulphides.

To imagine a kind of organism living all by itself, then, we are driven down to some very obscure bacterium like Thiobacillus, or perhaps to some of the simple algae that are able to utilize atmospheric nitrogen as well as atmospheric carbon dioxide. But these very bacteria and algae form, in nature, parts of complex systems of organisms which are mutually dependent in building or modifying series of compounds of these basic elements of nitrogen, carbon or sulphur. You can imagine Thiobacillus carrying on its sulphur operations in a biotic vacuum, but in fact it is thoroughly involved in the very complicated communal economy of soil organisms.

THE INTERRELATIONS OF ORGANISMS

Some kind of organism, sometime, must have crossed the bridge from the inorganic world to the world of living things all by itself. Some form of life must thus have existed alone for an eternity before, somehow, it began to diverge in function and thus become two kinds of life instead of one kind. To imagine something of this sort, on a basis of contemporary organisms, we are limited to things like the bacteria. But the bacteria today are integrated into the economy of organic nature as a whole, as dependent on the "higher plants" as the "higher plants" are on them; and the first pre-Cambrian protoplasmic ooze was probably something very different.

The point I am trying to make here is the interdependence of kinds of organisms in the world today. We have got into the habit of looking at the organic world as a mass of struggling, competing organisms, each trying to best the other for its place in the sun. But this competition, this "struggle," is a superficial thing, super-imposed on an essential mutual dependence. The basic theme in nature is cooperation rather than competition—a cooperation that has become so all-pervasive, so completely integrated, that it is difficult to untwine and follow out the separate strands.

The next step, after trying to think about an organism in iso-lation, might be to try to think about two organisms, and to imagine the possible different relationships that might exist between them. We at once come to a major dichotomy, whether the two organisms are of the same kind or different kinds. If they are of the same kind, they may cooperate to get food, like two ants, or fight over the possible food supply, like two dogs. The two individuals may co-operate for reproduction; or being of the same sex, may struggle for the opportunity to mate with a third individual; or still being of the same sex, may peacefully share reproductive functions with the third individual. If there is a parent-offspring relationship be-tween the two individuals, it may be protective, maternal; or it may be of the excluding, get-out-and-stand-on-your-own-feet variety.

If the two individuals are of different kinds, it is interesting to look for the direction of immediate benefit in their relationships. One may serve as food for the other. The direct benefit here flows in one direction. One may serve as protection or support for the other—the tree and the vine. In this case, "A" may derive benefit from "B" without harming "B"; or "A" may be a nuisance to "B" in varying degrees (depending on how heavily the tree becomes loaded with vines); or "A" may gain protection from "B" and give "B" a positive benefit in return. These cases of mutual benefit, of partnership, of symbiosis, are very numerous and take all sorts of forms, like the fungus and the alga uniting to make a lichen; or the protozoa in the intestine of a termite, digesting cellulose for the termite in return for protection. Or the numerous kinds of tropical trees that have developed special nectaries and cavities as home and food for stinging ants that in turn, in guarding their own nests, guard the tree.

It is probably useless even to try to catalogue the possible kinds of relationships between two different organisms, because we find that most of these relationships involve still other organisms; that the two compete to use some other kind of organism as food, or cooperate to support some other kind of organism, or utilize different parts of a third organism for food. In such an attempt to list the relationships between separate organisms, we come across the problems of the relationships within the community as a whole.

It is like trying to analyze the relationships between John, the barber, and Peter, the milkman. We soon get involved with Alfred, the grocer; with Peter's wife who works with Alfred's wife in arranging church suppers; with Elizabeth, who sells dresses imported from New York, which takes us out of the community and into ever more general and more indefinite, but none the less essential, relationships.

The logical basis for the examination of relationships among organisms is not individual organisms, or different kinds of organisms, but the biotic community. Here we encounter a problem in definition.

THE DEFINITION OF COMMUNITY

Most ecologists have worked in the temperate zone in pine forests or beech forests, and they have come to think of the biotic community as determined by a particular dominant kind of organism. The dominant is thought of as controlling the whole community either through sheer numbers of individuals, or bulk of protoplasm, or all-pervading biological effect. The whole assemblage of organisms living in a pine or beech forest is obviously dominated by the pines or beeches, so that these form a convenient guide to the limits of that particular community.

In the tropical forest, one looks in vain for any particular kind of organism that dominates in the sense that the pine or beech dominates its forest. My experience in the tropics has thus made me very dubious about the whole ecological idea of dominants. The idea works all right in the pine woods, but even in the temperate zone one must stretch the imagination to apply it in a lake or a stream or on the sea coast.

I think the essential element in the concept of the community is the interdependence of its various members to form a functioning unit. It is the next distinctive general organizational level above that of the individual and the population. Individuals of a given kind are organized, through their genetic relationships, into populations; but the behavior and history of these specific populations can only be understood in relation to the behavior and history of the other populations with which they occur. The community, it seems to me, might be defined as the smallest group of such populations that can be studied and understood as a more or less self-sufficient unit.

We thus arrive at a parallel with Toynbee's definition of an equally difficult concept, that of a "civilization." A biotic community might well be defined as "an intelligible field of study" from the point of view of the relations among different kinds of organisms.

It is a unit of ecological study, just as civilization is a unit of historical study.

Toynbee has pointed out that English history is in no sense intelligible if taken as a field in itself. From its first clear chapter, the conversion of its people to Christianity, to its last, the establishment of industrial economy, the course of events in England fails to make sense except through the consideration of events in neighboring national states. But in tracing the genesis of these events, the student finds that it is not necessary to cover the whole world as a field, but that more or less definite limits in time and space are reached, limits which Toynbee in the present instance takes to define the particular field of Western Christian civilization. Within this field, the majority of relationships are directed inward, as they are in his other contemporaneous civilizations, the Orthodox Christian, the Islamic, the Hindu, and the Far-Eastern.

A concept like that of the "intelligible field of study" does not give a sharply defined series of units: but the material to be studied, at this level, is not composed of sharply defined units. Human culture is a continuum, but it is a continuum with pronounced modes both in time and in space, and the problem is to recognize these modes and to trace their development and characteristics. Even such an apparently isolated phenomenon as the Mayan civilization is related to the general nexus of culture. It must have started with people who had brought certain cultural equipment with them when they parted from the Old World along the bridge of the Aleutian Islands or across the Pacific, and this cultural equipment would be the common relationship between the civilizations developed in the Old and New Worlds. Among Old World civilizations, the relationships are, of course, numerous, in part obvious, and of a variety of sorts. The important thing is that these relationships are external, that is, their study is not essential to an understanding of the development of events within the civilization.

It is not particularly strange that civilization and biotic communities should be subject to the same type of definition process, since both are biological phenomena, involving the aggregation of individual organisms. There would be even less difficulty in making the analogy if Toynbee had been concerned with "cultures" instead of with "civilizations," which are essentially a particular kind of cultural development. Civilizations happen to be the only cultural groupings that have got much involved with the special problem of history. If one were studying human ethnology, rather than history, interest would focus on the development of cultural relationships in general, and civilization would fall into perspective as a particular type of development shown by a few scattered cultures.

THE LIMITS OF THE COMMUNITY

An oak tree may seem a world in itself. It may have hundreds of organisms of dozens of kinds that depend on it for food; it may give support to Spanish moss and other epiphytes; protection to birds that nest in its branches. The leaves that it drops decay, and the characteristics of the litter of oak leaves may determine the kind of soil organisms, and the direction of soil development.

But this complex of organisms does not form a community. It is not an intelligible unit of study. If we start to analyze the relationships of the various organisms associated with the oak tree, we find ourselves constantly led away from that association into some larger grouping. The caterpillars that feed on the oak leaves are developmental stages of a butterfly that gets food from the goldenrod in a clearing. The birds nesting in the oak tree forage through the nearby forest, and their behavior cannot be understood in terms of the tree, but only in terms of the forest. Similarly, we find that the association of organisms in the soil depends not so much on the fact that they are under an oak tree, as on being part of the oak forest.

Our biotic community, then, is the forest, not the association in the tree. The complex of organisms making up the forest does form an intelligible field of study, where the events and relationships can be understood with only casual and occasional reference to external phenomena. The range of the birds nesting in our particular tree depends on their relations with birds in neighboring trees, on the "territory" in the forest that is available for each family. The kind of food that they are able to gather depends on a selection from the kinds available in the forest. The number of caterpillars of a given kind eating the leaves of the oak tree depends on the density of population of that particular species in the forest, which is a matter of balance of parasites, climatic relations, and so forth. The characteristics of the forest as a whole can be related to the regional climate, to the physical and chemical factors of the general environment. The climate within the forest can be understood in relation to the density and size of the trees. The forest constitutes a biotic community.

A pool in our forest may, or may not, be inhabited by a distinct community: the answer would soon become obvious in the course of study of the organisms in the pool. If the pool formed an intelligible unit without special reference to the forest, it would certainly form a separate community. But we might find that the behavior of the inhabitants of the pool was constantly conditioned by the character of the forest. The kind of plants growing in the pool might depend on the shade relations with the oaks growing around the margin. A good proportion of the animals in the pool would probably turn out to be insect larvae, and the kind of insects in the particular pool might depend on whether the adults were able to fit into the conditions in the surrounding forest. The acidity of the water in the pool might depend on the soil character of the forest and on organic matter washed into it from the forest in rains. If, in studying our pool, we found that we had thus constantly to refer back to the forest, we could hardly consider it as more than a special sort of "niche" in the general forest community.

On the other hand, a small lake would probably turn out to form an intelligible field of study in itself. The naturalist studying the lake could likely get along very well with only occasional and casual reference to the communities inhabiting its shores. To understand the physical environment of the lake, he would have to study its mode of origin, whether from glacial action, from uplift across a stream drainage system, or from some more remote sequence of geological events. He would be concerned with the depth of the lake and with the regional climate and its relation to temperatures at various depths. The kind of fish populations inhabiting the lake and their density would depend on host-prey relations among these populations, on the sort of breeding places available in marginal vegetation, and their extent. Even the larval insects found in the vegetation at the lake margins would propably depend more on the characteristics of the lake community than on the characteristic of the forest community adjoining the lake.

If the lake were big enough, we might find that it was inhabited by a series of communities which formed intelligible fields of study in themselves. The problem of study of great lakes thus approaches in nature the problem of studying life in the sea. The lake takes on the character of a geographical region, inhabited by many kinds of communities, tied together chiefly by factors of the regional climate, and the regional geological history.

A biotic community may form an intelligible field of study, but it is never a completely isolated field of study. The birds that nest in the oak community may spend the winter in South America. Large mammals, especially predators such as the puma and jaguar, range through all kinds of communities, although the activities of a particular individual are usually confined to a rather limited area, perhaps a single community. The fluctuations in density of a particular species of organism in a given community may depend on factors that affect the population in the whole geographical region.

SUCCESSION: THE HISTORY OF COMMUNITIES

Nor are communities stable in time. Charles Elton has described this very well in a passage in his book on *Animal Ecology*:

"If it were possible for an ecologist to go up in a balloon and stay there for several hundred years quietly observing the countryside below him, he would no doubt notice a number of curious things before he died, but above all he would notice that the zones of vegetation appeared to be moving about slowly and deliberately in different directions. The plants around the edges of a pond would be seen marching inwards toward the center until no trace was left of what had once been pieces of standing water in a field. Woods might be seen advancing over grassland or heaths, always preceded by a vanguard of shrubs and smaller trees, or in other places they might be retreating; and he might see even from that height a faint brown scar marking the warren inhabited by the rabbits which were bringing this about. Again and again fires would devastate part of the country, low-lying areas would be flooded, or pieces of water dried up, and in every case it would take a good many years for the vegetation to reach its former state. Although bare areas would constantly be formed through various agencies, only a short time would elapse before they were clothed with plants once more."

If an area of forest is cleared, a crop planted, and the land then abandoned, this land will support a succession of plant communities of different types which will form an orderly sequence until finally the stable, terminal forest community again forms. The kinds of communities in the sequence depend on the general climate of the region and the specific soil and drainage characteristics of the cleared area. We are all familiar with the obvious stages in succession called "second growth," but some rather permanent-looking types of vegetation are essentially temporary. The pine forests of

Florida, for instance, are maintained by the periodic burning which the pines withstand, but which kills the oaks. If such an area is long protected from burning, it slowly changes from a pine forest to an oak forest.

Ecologists in North America have been much preoccupied with the description of this succession process, and have subjected the phenomenon to elaborate study. This is useful, because such information is necessary for the proper management of our land resources; but it seems to me that these students have put an emphasis on succession that is out of all proportion to its general role in natural history.

This emphasis is understandable in North America because it is a continent that has been devastated by man, and my own indifference to succession undoubtedly stems from the fact that I have worked in areas as remote as possible from human interference. The effect of man on the landscape is profound and far-reaching. He must form the most powerful geological force at work on the surface of the earth today. The effects of floods, of hurricanes, of earthquakes, of volcanic irruptions, of natural erosion, the effect of all of these processes fades into insignificance beside the effect of man with an axe.

The man with the axe is an integral part of nature, and the consequences of his activities make an interesting and important, though dismal, field of study. But he is, geologically, a new sort of phenomenon, and study of his activities thus has only a limited usefulness in the attempt to reach an understanding of how biological processes came to be the way they are.

The effect of man on the biota of a region is really profound. This has been brought home to me because for many years, in the Villavicencio region of Colombia, I have alternated between the neighborhood of the town, where man is firmly established, and regions where he is still only a casual intruder. The common plants, the weeds, the roadside shrubs, the second-growth trees, where man has settled, are the rare plants in the areas where he has not yet penetrated. The common roadside butterflies, whose larvae feed on these weeds, are the scarcest of insects in the untouched areas. The balance of the whole fauna and flora, from the bacteria and protozoa of the soil to the trees and rodents, the snakes and predatory cats, has been changed.

Succession is a common and natural phenomenon, but man makes it universal in the areas where he settles. In this upper Orinoco country, where man is a rare animal, the weeds and shrubs and trees so common in the settled areas are found only on new sandbars in the rivers; in spots in the forest when a giant tree, felled by lightning, has pulled its neighbors down with it, and made a little opening in the forest; on mountainsides, where a slide of rocks has laid a surface bare. Succesion, in an untouched area, is limited to situations like these.

Natural succession, as distinguished from the man-made variety, may be a striking and geologically rapid phenomenon in some sorts of places, especially along beaches, in sand dune country, and in fresh-water environments. There must also have been a succession of communities everywhere, through geological time. These changes were drastic and relatively rapid in some places and times, as in the temperate zone during Pleistocene glaciation; but they have surely been exceedingly slow in other areas, such as midcontinental South America.

THE HABITAT AND THE NICHE

"Community" and "succession" are common words in the ecological vocabulary. Equally common is "habitat". The habitat of an organism is the place where it occurs. It might thus be defined as the environment of a particular kind of organism. The habitat may be co-extensive with the community, may (rarely) include several types of communities, or may represent a niche in the community. The description of the habitat would include the physical and chemical environment as well as the biological environment (communal relationships).

Generally, the habitat of an organism corresponds to a particular niche in the community. I am using "niche" here in a slightly different sense from that usual among ecologists, but since I have diverged widely from custom in defining the community, perhaps I may also use the word niche in an unorthodox way.

Niche is customarily used "to describe the status of an animal in its community, to indicate what it is *doing* and not merely what it looks like" (Elton). The trouble with that, for me, is that niche has connotations of place—the dictionary says "a place, condition, or the like, suitable for a person or thing". I would prefer to use "role" to indicate what an organism does, and "niche" to indicate its physical place in the community structure. In a forest community, the organisms living free in the tree canopy would live in a niche; those ranging over the ground (what I like to call the forest floor zone) another niche; those among the litter of dead leaves, still another; those in the soil below the forest, still another niche.

There may be endless special niches. Tracing them out in a tropical forest is an especially fascinating game, and the multiplicity of niches in such a forest makes possible the diversity of kinds of organisms that, in their sum, make up the community. Thus the epiphytic plants (orchids, ferns, many cacti, air plants, and so forth) constitute a special niche of plants that get access to the light of the upper forest zones by perching on the branches of trees. The bromeliads or air plants, the family to which the pineapple belongs, are conspicuous members of this epiphytic niche, and they in themselves form yet another niche.

The bromeliads mostly have long narrow leaves that grow from a common center, where the bases of the leaves form a watertight vessel, which the botanists aptly call the "tank". A big bromeliad may have a lot of water—sometimes several quarts, though the average would be a pint or less. This water collects debris of all sorts, and becomes a quite rich infusion which serves as a source of food for the bromeliad, but it also in itself becomes the habitat of a complex fauna. The water in these bromeliads contains, besides a host of microscopic organisms, many kinds of insect larvae such as those of mosquitoes and damsel-flies, special worms and snails, and even the tadpoles of certain kinds of tree-frogs. This complex of organisms does not form a community because the relations within the bromeliad can only be understood in terms of the forest as a whole. A major proportion of the inhabitants of the niche are larval forms of organisms whose adults live free in the forest canopy; the bromeliad itself is an organism living in the canopy zone, and the physical characteristics of the water environment in the bromeliad tank are products of the forest climate.

ROLES IN THE BIOTIC COMMUNITY

Elton's concept of niche is more directly concerned with the kind of activity of the organism in the community, with its role. Thus among the animals in a community, there is a whole series of predators, ranging from very small predators that feed on tiny organisms to big predators that feed on big animals. There is a similarly graduated series of herbivores. As Elton points out, diverse animals may play closely similar roles in different kinds of communities in different parts of the world: in many different communities there is some species of large snake that preys exclusively on other snakes; some species of bird that specializes in picking ticks off large herbivores, and so forth.

These various roles are chiefly defined in terms of food, and most studies of the interrelations of animal populations in communities have been dedicated to the description of food-chains. The classical food-chain of the fleas has perhaps been best stated by some anonymous author:

"Great fleas have little fleas upon their back to bite 'em,
And little fleas have lesser fleas, and so *ad infinitum*.

The great fleas themselves in turn have greater fleas to go on,
While these again have greater still, and greater still, and
    so on."

Working out food-chains can become a fascinating occupation—one that still has not had anything like the attention that it
deserves—but the resulting diagram is apt to give too rigid an
impression of the animal relationships. Such chains start with some
plant, or some type of vegetation, and begin the animal chain with
the smallest kinds of animals that feed on the vegetation—what
Elton calls "key industry animals" because they are apt to be present in enormous numbers and furnish food for a great number of
predators. A very simple chain of this sort could be: grass—grasshoppers—frogs—snakes—hawks. But one could make the shorter
chain of grass—deer—puma. And one could leave the long chain
at any point by going to internal parasites, or smaller fleas, instead
of to bigger predators.

The game of building food-chains is only practical with the
animal members of the community. Yet the community may include
also hundreds of different kinds of plants, existing in some sort of
population balance that obviously doesn't depend on which plant
eats which. The food-chains are complicated enough, but they are
still too simple a method of expressing community relationships.
They stress the competitive aspects of the community and I suspect
that the stability of biotic communities depends on cooperative
relationships at least as much as on competitive relationships.

In a food-chain the end is some master predator, big enough
and strong enough to have no enemy in turn. This seems the lord of
the community, as the lion is thought of as the king of beasts. Yet
these top predators occupy about the least important role. The lion
isn't the king of the community. He is found in a sort of community
that is rich enough or luxuriant enough to be able to afford a few
master predators.

The study of communities is complicated by the fact that it is not easily attacked by the experimental method. It is difficult to remove pieces of a community one at a time to see what happens, and only the very simplest sorts of communities can be maintained under laboratory control—communities that are perhaps too simple to serve as models for the relations under complex natural conditions. We are forced, then, to depend rather largely on observation and inference in the study of communal relationships, though there is considerable room for the development of the experimental method in studying individual kinds of relationships.

In the next two chapters, I shall attempt some analysis of particular communal relationships. The present chapter started out with emphasis on the inter-dependence of organisms, which leads to the development of a community sort of organization, and in the next chapter I shall attempt to document this with a variety of instances and kinds of mutually helpful relationships that have been described among organisms. Then, in the following chapter, I shall try to make a rather general description of a single kind of relationship that is normally considered as antagonistic, the relationship of parasitism. After that we shall shift emphasis again from the community to the individual and to organization in the form of populations, treating the individuals and populations this time from the point of view of behavior.

CHAPTER IX

Partnership and Cooperation

SCIENTISTS and philosophers have long been impressed with the competitive, the antagonistic relations among organisms. These are more obvious, more striking than the cooperative relations, which result in the mutual dependence among the members of the biotic community described in the last chapter. The competitive aspects of communal relations have been phrased in easily remembered slogans like the "struggle for existence" and the "jungle law of tooth and claw." The cooperative aspects of communal life seem to be less susceptible to apt and catchy phrasing.

I wonder whether this is related to some general character of the thought process. We seem generally to be more impressed with strife than with cooperation, whether in nature or within our own species. A war anywhere, any time, rates the biggest headline type available, while a treaty for reciprocal trade has scarcely any news value. A street corner fight is always interesting, but a boy scout helping a blind man attracts no crowd. There are, of course, exceptions, and cooperation can be dramatized, but it takes effort and peculiar circumstances.

The preoccupation of naturalists with predation, parasitism, food-chains and the struggle for existence may then merely reflect a general human love of gore. But this emphasis in biology has, I think, had unfortunate consequences because it has served as a basis for certain trends in social philosophy. The "survival of the fittest" can be used in defense of war, if you grant that the fightingest are are the fittest. The "struggle for existence" can be used as an explanation for the prevalence of poverty and disease, and for the futility of attempting remedial action.

Actually, I think there has been an increasing tendency in biological writing to stress the cooperative rather than the competitive aspects of relations among various kinds of organisms. But this tendency has not been adequately reflected in the thinking of the social philosophers, who have tended to confine their biological explorations to the post-Darwinian, or Thomas Huxley period. I don't think the social philosophers are entirely to be blamed for this. The biologists have failed to make their growing knowledge, their accumulating facts and concepts, easily available to the philosophers. Perhaps it would require another Thomas Huxley to integrate modern biology and philosophy. Of the Huxleys, Julian is a better biologist than his grandfather, but he seems to have been much less successful in communicating biological ideas to the social philosophers of the contemporary scene.

But I had better get back to the subject at hand, which involves relations among organisms, rather than relations between biologists and philosophers. It is actually unsound, I think, to try to sort out relations among organisms into beneficial and injurious categories. The biotic community, as I tried to show in the last chapter, is a functioning unit, and the various component populations serve to build up the unit as a whole. Stressing competition among these populations is, in a way, like stressing competition between muscle tissue and brain tissue for food within an individual organism.

A caterpillar is considered to be a parasite of the tree that it lives on. The ovaries, though, are not usually classed as parasites of the stomach, nor the stomach as a parasite of the mouth. Yet the relations between caterpillar and tree and ovaries and stomach are analogous, particularly if the tree depends for the fertilization of its flowers on the butterflies produced by the caterpillars. The relations among cells within a tissue, tissues within an organism, and organisms within a community, are directly comparable, representing different levels of organization.

THE PROBLEM OF LEVELS

We are concerned often in science with the problem of levels of organization. The shifts in level are not always clean-cut, like steps, though neither are they arbitrary divisions in a smooth slope. The levels that concern biology have been given convenient names as the molecule, the cell, the individual, the community. The final biological level, I suppose, would be the biosphere as a whole, the complex of life processes over the surface of the earth, from which the next jump would be to the organizational levels of the astronomers. At the other extreme the atomic levels of the physicists lie below the molecular level of the biochemists.

There are various difficulties with this neat series. For one thing, it is possible to climb from the atomic level of the physicist to the cosmic level of the astronomer by steps that do not lead through the biological staircase. As naturalists we might do well to leave the atomic and cosmic ends of our organizational scale strictly alone, lest we get dizzy. We are on safer ground in the area between the cell and the biosphere.

The individual, with its cells, tissues and organs, represents a compact, tight, definite organization, as compared with the loose, vague and indefinitely organized community. But both levels of organization are parallel characteristics of the life process, designed for the apparent purpose of insuring the perpetuation and fullest development of that process. Though we had better be circumspect in flirting with that word "purpose"—keeping a firm grip on the defensive adjective "apparent."

To get back to the communal level, which is our concern here: I think all intracommunal relationships must have a cooperative aspect, else the community would be unable to function. Even the prey-predator, the mouse-cat, relationship is cooperatively adjusted. The mice have a high reproductive rate, which might be taken as an adaptation necessary to produce plenty of cat food; the cats a lower rate, necessary to avoid using up the supply of mice. Without control by cats (among other factors), the mouse population might rapidly reach a density harmful to the best interests of the mice. With an abnormally dense mouse population, the mouse food supply becomes uncertain, making conditions precarious for all mice; territorial squabbling sets in; and repercussions occur all through the biotic community. The mice thus benefit from the cats by way of population control; while the cats benefit from the mice, by way of food supply.

This is mutual benefit—and however absurd my reasoning to mutual benefit may seem, its absurdity stems only from the fact that we are conditioned to consider mice and cats as enemies, thinking of them as isolated populations of two kinds of animals, not as functioning elements of a biotic community.

There is a whole series of words for particular kinds of relations among organisms, words like symbiosis, commensalism, mutualism, parasitism, predation. Attempts at definition of these words have not been notably successful, probably because the different relations within communities do not fall into distinct, clean-cut categories susceptible of logical arrangement. But we need some frame on which to hang our discussion, and these words furnish as convenient a structure as any.

THE CONCEPT OF SYMBIOSIS

The first word, symbiosis, is perhaps the most difficult of all because each biologist has given it a somewhat different meaning. The word was coined in 1879 by Heinrich Anton de Bary, a German botanist who was a pioneer in the study of fungi. The word is based on two Greek roots which mean, simply, living together, and de Bary used it in a general sense for any close association between two different types of organisms, whether the association was obviously of mutual benefit, or whether one organism was living entirely at the expense of the other.

De Bary was familiar with both types of relationship, since he was one of the first to appreciate the mutual advantages of the association of fungus and alga to make a lichen, and since he was also among the first to demonstrate infection—the parasitism of one organism by another causing physiological change manifest as disease. For de Bary, the association of fungus and alga to make a lichen, and the association of fungus and potato plant to make potato blight, were both examples of symbiosis. One would be mutual, or reciprocal, symbiosis; the other antagonistic symbiosis, or parasitism.

The American Society of Parasitologists, with due respect for origins and tradition, has endorsed this general meaning for symbiosis. But I suspect that it is a lost cause, like most of the causes of the purists. Symbiosis makes a very convenient opposite for parasitism, and it has gained wide currency in this special sense. When we have an ancient and well established concept like parasitism, it is difficult to see the gain in trying to substitute "antagonistic symbiosis." On the other hand, "reciprocal symbiosis" is redundant if the single word "symbiosis" can be made to do the work. So, with due apologies to the parasitologists, we shall consider symbiosis and parasitism to be separate words, with opposite meanings. The protozoa that digest cellulose in the intestines of termites represent symbiosis in this sense. The amoebae that disrupt the intestinal wall and cause disease in man represent parasitism. This may violate tradition, but it is very convenient, and corresponds to wide usage.

Symbiosis, in this narrow sense, represents the most intimate of the mutually beneficial relationships that may exist between different types of organisms. The partnership is so close that usually neither organism can live without the other, though in some cases one or the other of the partners may also be able to lead an independent existence.

ALGAE AND FUNGI IN PARTNERSHIPS

The lichens were the first example of symbiosis to be thoroughly studied, and they remain one of the most striking examples. The lichen, as I pointed out in Chapter III, is composed of a fungus and an alga living in close association: each kind of lichen being composed of a particular kind of fungus and a particular kind of alga. The fungus provides a structure in which the algae may be held, and gathers water and inorganic salts. The algal cells, which contain chlorophyl, are able to synthesize starch through photosynthesis, building up food materials for the fungus. The association of the two is so intimate that they appear to form a single, discrete "plant."

Both algae and fungi are involved in a surprising number of symbiotic relationships with many other kinds of organisms, aside from the lichen case in which they are symbiotic with each other. Organisms which do not themselves have chlorophyl have frequently discovered the advantage of symbiotic relationships with algae. Algae have been described as living in symbiosis with protozoa, fresh water sponges, Hydra, reef corals, sea anemones and various other kinds of animals.

Symbiosis of other organisms with fungi is even more widespread than symbiosis with algae, and more important in the economy of nature. This is because a great many different kinds of seed plants have come to depend on the help of fungi in absorbing nutrient materials through their root systems. The association between the root of a seed plant and a fungus is called a mycorhiza (fungus root). In some cases, it looks as though all of the advantage accrued to the seed plant, which would thus be a sort of parasite on the fungus; while in other cases the advantage seems mostly to lie with the fungus. Whatever the balance in individual cases of partnership, the whole phenomenon is probably best considered as an arrangement for mutual benefit, a symbiotic relationship.

Mycorhizae have been found in practically all of the families of seed plants. The fine thread-like hyphae of the fungi are more efficient agents for absorbing nutrient materials from the soil than are the root-hairs of the plants; and the fungi in turn, lacking chlorophyl and the ability to build up carbon compounds themselves, find the seed plants convenient partners. Pine trees, for instance, are usually dependent on mycorhizae, and are able to grow without their corresponding fungi only in exceptionally fertile soils. Efforts at reforestation with pines have failed in soils where the symbiotic fungus was lacking, until proper fungi were introduced.

The orchids provide the classical examples of mycorhizal symbiosis. The phenomenon is relatively easily studied in these plants because of the horticultural practice of germinating orchid seeds on gelatin media—the same sort of technique that would be used, anyway, for studying fungi. Most orchid seeds fail to germinate if the proper fungus is not present, though the seeds may be induced to sprout if chemical conditions in the agar medium are just right. It is like getting the proper formula for bottle-feeding the baby. The orchid root, without its fungus, is incapable of gathering up enough food materials. A few orchids have no chlorophyl, and could be regarded as completely parasitic on their fungus, which in turn lives on humus and other decaying vegetation. This is a sort of turnabout, since fungi in general are pretty expert at parasitizing other organisms themselves.

SYMBIOTIC RELATIONS OF BACTERIA

The bacteria have become more deeply involved in symbiosis than either the algae or the fungi. The best known partnership between multicellular organisms and bacteria is probably the association between plants of the bean family and the nitrogen-fixing bacteria that live in their root nodules. The role of bacteria in fixing nitrogen was established in the last years of the nineteenth century by several different workers, and the process has been the subject of intensive study ever since, though we are still far from a complete understanding of its chemistry and physiology.

The bacteria associated with legumes can be grown in "pure culture" in the laboratory, but they do not fix nitrogen when they are not associated with the proper bean plant. Furthermore, a given species or strain of bacterium is able to fix nitrogen only in association with one or a few particular species of beans: the same phenomenon of host specificity that we find with parasites. A given strain of bacterium will sometimes invade the roots of various bean species and form nodules, but careful measurements will show that no nitrates are produced. Where the partnership is chemically successful, the bean plant is stimulated by the bacterium to produce a red pigment, very similar to if not identical with the haemoglobin of mammalian blood. We do not yet know exactly what role this pigment plays in the chemical process of building up soluble nitrates from the nitrogen gas of the air, but it serves as a clear marker of the cooperative nature of the process. The bean plant is not merely a convenient refuge for the bacteria: it is also an active partner in the chemical activity to which the bacteria are dedicated.

The association between the bacteria of the genus Rhizobium and the plants of the legume family is a striking example of symbiosis, but it is only one item in the long list of cooperative bacterial activities. Any attempt to catalogue all these activities here would be out of the question; I can only quote a few examples more or less randomly selected.

Symbiosis between bacteria and insects should be mentioned, because it is the subject of a voluminous literature. The association between insects and bacteria is so general and so close that it must be important, though we know little about its nature. Bacteria have been found living within the cells of almost all kinds of insects, sometimes in special organs called "mycetomes" which seem to exist simply for the purpose of harboring bacteria. Attempts to grow the bacteria in culture media, separate from the insect, have failed; and where it has been possible by some special treatment to kill the bacteria without killing the insect, the vitality of the insect has been found to be greatly impaired. The two kinds of organisms are thus mutually dependent and the relationship must be symbiotic.

We know that in some cases bacteria that live in the intestines of other organisms produce food elements such as vitamins necessary for the host organism, that it cannot build up for itself and that are not present in adequate amounts in available food. The intracellular bacteria of insects may also play some little understood nutritional role.

Bacterial symbionts may aid in the digestion of particular food materials that cannot be broken down by the digestive enzymes of the host organism. Thus the larva of the olive fly maintains a supply of special bacteria in pockets of the oesophagus which apparently help in the digestion of the olive juices. These bacteria are carried from one generation to the next through organs in the oviduct of the parent fly, which insure that each egg will be smeared with bacteria as it is laid. There are a variety of such elaborate arrangements in insects for insuring that the symbiotic bacteria will be passed on to the offspring, strong evidence of the utility of the bacteria to the insects.

THE DIGESTIVE PROBLEMS OF INSECTS

Insects have come to depend on organisms other than bacteria, such as yeasts and protozoa, for help in the problem of digesting special materials. The association between termites and protozoa is a good example. Termites are an ever present element in the tropical fauna and form a sufficiently fascinating field of study in themselves, because of their highly developed social organization and

their varied adaptations for getting at the dead wood on which they
live. They are delicate, white, soft-bodied insects that live always
within the shelter of their nests, sometimes building long covered
runways from those nests to inaccessible bits of wood, so that they
can carry on their activities without ever exposing their fragile
bodies to the hazards of the open air.

Now wood is mostly cellulose, a notably difficult substance for
animals to digest. Mostly, in fact, cellulose cannot be utilized by
animals until it has been broken down by bacterial action. The
termites, however, have solved their cellulose problem by cooperation with protozoa. Each kind of termite has its particular kind,
or kinds, of protozoa. The termites and these intestinal flagellates
show parallel divisions into genera and species, indicating an ancient association, with parallel evolution in the two partners. Various ways have been worked out of killing the protozoa in termites
without injuring the termites; but without their protozoa, the termites die of starvation within a few days or weeks. The protozoa
have also come to be completely dependent on the termites, dying
very rapidly if removed from their hosts. The protozoa in a termite
may be very numerous, representing from 16 to almost 50 per cent
of the total weight of the termite.

SYMBIOSIS AND PARASITISM

These various examples—the fungus and alga forming a
lichen; fungus and seed-plant root forming a mycorhiza; bacteria
and legumes associating for the fixation of nitrogen; bacteria, yeasts
and protozoa aiding insects in food digestion—serve to indicate the
general nature of the partnership relations that can conveniently be
included under the general term of symbiosis. The association in all
cases is an intimate one between two or more very different kinds
of organisms: usually between a large, complex organism and a
large number of microorganisms. I think the word could be well
restricted to this intimate association within a single body, so that

the similarities and contrasts with parasitism—a comparably inti-mate association in which the benefit is not obviously or immedi-ately reciprocal—can be brought out. The line between symbiosis and parasitism, in this sense, is not easy to draw, but the implications of this are best left for discussion in connection with parasitism, in the next chapter.

Symbiosis, in this restricted sense, is only a very special class of the multiform mutually beneficial relations that have developed among organisms. It is spectacular because the two partners are so obviously and completely dependent on each other, but it is hardly more important or real than the mutual dependence of organisms that may appear to have only casual or indirect contact with each other in the general course of community life.

THE MUTUAL DEPENDENCE OF ANIMALS AND PLANTS

We might jump from the one extreme of intimate association in symbiosis to the other extreme of general and indirect coopera-tive relations, before giving a few examples of intermediate sorts of interdependence. The mutual dependence of plants and animals is perhaps the most general of these cooperative relations.

The prime characteristic of the higher plants is their ability to synthesize starches with carbon dioxide and water. The prime char-acteristic of the higher animals is their mobility. The mutual rela-tions between higher plants and animals depend on the sharing of these two characteristics—the animals depending on the plants for food, and the plants on the animals for fertilization and dispersal.

Insect pollination of seed plants is so general, and involves so many complex adaptations on the part of both the plants and the insects, that even without fossil evidence one would assume that the evolution of the two groups had been parallel. The insects developed first, and it is easy to believe that the evolution of most of the groups of seed plants has depended on the presence of an animal group such as the insects. The mobility of the insects has solved the plant problem of how to bring the gametes together in the terrestrial environment. Wind pollination, looking at the seed plants as a whole, is a much less successful, much less general, solution of the problem. It requires the liberation of enormous amounts of pollen, and seems to be most frequent in plants that occur in fairly pure stands (conifers, grasses). It is difficult to imagine the development of a complex plant community like a tropical rain forest if the distribution of gametes were dependent on a random mechanism like wind.

There is no use in cataloguing the endless adaptations of flowers to attract insects, or of insects to get at the nectar offered by the flowers. Butterflies, flies and bees and all of the colorful, conspicuous and fragrant flowers, represent a maze of mutual adaptations. It is a fascinating field of investigation that has attracted many students, including Charles Darwin, who wrote a book entitled *The Various Contrivances by which Orchids are Fertilized by Insects.*

We might cite one extreme case of mutual adaptation between an insect and a seed plant: the case of the Yucca moth (Pronuba) and the Spanish Bayonet (Yucca) of the southern United States. A particular species of Pronuba is associated with each species of Yucca. The moth emerges at the time of the opening of the Yucca flowers, which may be for only a single night. It rolls together a ball of pollen from the flowers of one plant, flies to another where it makes an opening into the pistil to lay four or five eggs, inserting the pollen ball into this opening. The plant is completely dependent on this particular moth for fertilization, and the moth on this particular kind of plant for larval food.

Plants also have come to depend greatly on animals for dispersal. Witness all of the varieties of burrs and sticktight seeds; and the diversity of fleshy fruits. Such fruits are of no use to the plant except insofar as they serve to induce some animal to eat them, and then scatter the seeds in his faeces.

Even the nutritional relations of plants and animals are not all one way, from the plant to the animal. The value of manure is not limited to gardens; in some types of biotic communities it may be an important element of plant food. Probably more important in general, though, is the activity of animals in modifying soil conditions through their burrowing activities. This again served Darwin as the subject of a book—*The Formation of Vegetable Mould through the Action of Worms.*

COOPERATION IN THE INSECT WORLD

Between these extremes of the close mutual association of organisms in symbiosis, and the general interdependence of the plant and animal members of a biotic community, lie a host of reciprocal relationships that are usually described under some general term such as "mutualism" or "commensalism". I have not seen any general review of these associations, though I should think they would form an interesting and instructive subject for a book.

The bizarre world of the insects has many such mutual associations. Social insects, such as the ants, are particularly apt to be involved. We have the ants attending their aphid "cows", pasturing them out from plant to plant in return for the drops of honey secreted by the titillated aphid. Then there are the endless associations between ants and special plants in the tropics: Acacias with their great hollow bull thorns inhabited by stinging ants that more than reenforce the defensive power of the thorns; trees like Triplaris and Cecropia that provide special hollow compartments in their trunks and stems for particular kinds of militantly vicious ants; the leaf nectaries provided by these and other plants as a food source for their ant protectors.

The agricultural enterprises of the leaf-cutting ants of the genus Atta might well also be considered as mutually advantageous to the ant and the fungus. The ant eats the fungus, and the fungus in turn gets the very best treatment under ideal environmental conditions, and thus probably achieves a more luxuriant growth and wider dispersal than would otherwise be possible. The association is strikingly similar to that between man and his cultivated plants. The ants build great nests with elaborate gallery systems, where they store masses of bits of cut leaves on which they grow the fungus which they use as food. I think this economy was first described by Thomas Belt in his book, *The Naturalist in Nicaragua,* and Belt's account remains among the best.

THE ANT "GUESTS."

These clearly mutual relations between ants and plants and other insects verge into the less clear case of the inquilines—the various guests found in ant nests. William Morton Wheeler, the great American ant student, estimated in 1910 that 1500 different kinds of ant guests had been described, and that this number would be more than doubled when these insects were more thoroughly studied in the tropical regions. Wheeler remarks that the diversity of the relations of these inquilines to the ants is even more extraordinary than their numbers. They form, he says, a "perplexing assemblage of assassins, scavengers, satellites, guests, commensals and parasites, and the same species may assume different relations towards the ants in its different developmental stages, or may be sufficiently versatile to combine the habits of different groups."

Wheeler divides the inquilines into four groups, depending on the attitude of the ants: those that are persecuted as intruders, and that must elude the ants to get at their food; those that are indifferently tolerated by the ants, living in the nests without being noticed or arousing any great animosity; those that are amicably treated, licked, fondled, fed and even reared by the ants; and those that live as external or internal parasites on the ants, with the characteristics of parasitism in general.

The third group, those that are amicably treated by the ants, are of most interest to us in the present connection. They are mostly beetles, but include representatives of several different families.

They have many peculiarities in common: they are generally of a peculiar reddish color; they usually bear tufts of red or golden hairs that are assiduously licked by the ants; their mouthparts are modified, since they have come to depend on food given them by the ants by regurgitation; and the antennae have become greatly modified, sometimes for stroking the ants, sometimes toughened to serve as handles whereby the ants can drag the inquilines about.

The great Dutch ant student, Erich Wasmann, was so impressed by the peculiarities of these inquilines that he suggested that they resulted from a special evolutionary process which he called "amical selection", in contrast to Darwin's "natural selection". He compared the inquiline evolution with the evolution of man's domestic animals, believing that the ants themselves act as the selecting agency, not only rearing and feeding the inquilines, but actually producing, by a kind of unconscious cultivation, such characters as the hair tufts and the modified antennae.

It seems improbable that the inquilines have been involved in any special evolutionary process, and Wasmann's ideas have received scant attention. But certainly these insects and their relationships with ants form a field of study that warrants more attention than it has so far received.

The ants and other social insects, with their partners, guests, hangers-on and parasites, represent a special problem in biotic relations, closely analogous to man with his partners, guests, hangers-on and parasites. The relations of these social animals, however, serve to underline the gradualness and tenuousness of the division between symbiosis and parasitism: the difficulty of drawing sharp distinctions between associations for mutual benefit; for the benefit of one partner without harm to the other; and for the benefit of one partner to the detriment of the other. Establishing the benefit or the detriment is not always easy—since often it depends on your guess or my guess as to whether a certain number of fleas are good for the dog or not. Which involves the definition of parasitism.

CHAPTER X

Parasitism

A PARASITE, in the broadest sense, is any organism that lives at the expense of another. Yet when we use the word, we usually have reference to a much more special sort of phenomenon, to an intimate association between two different kinds of organisms whereby one has come to depend completely on the other for sustenance. We think of a small organism living in or on the body of a large organism, usually causing more or less appreciable injury to the large organism (the host) which may be manifest as disease, but not necessarily causing the death of the host. In fact, the most highly developed form of parasitic relationship is supposed to cause a minimum of injury to the host, since the maintenance of the host in good condition is obviously to the "best interests" of the parasite.

PARASITISM VERSUS PREDATION

Parasitism thus differs from symbiosis in the one-way direction of the immediate benefit in the partnership. It differs from predation in the mechanism of exploitation. The predator catches, kills and eats the prey. A single predator requires an abundant supply of prey to serve as food, and the predator is in general bigger than the prey. Many parasites may live at the expense of a single host,

137

so that the numerical as well as the size relations are the reverse of those of the predator-prey combination.

Predation is typically an animal activity, though there are a few predatory plants, like Venus's flytrap. These plants must depend on devices to lure the prey within reach, where it can be trapped, since they lack the ability to chase their prey. Similar tactics have been adopted by various predatory animals, which have found it easier to lure the prey than to chase it.

Predation is thus a widespread phenomenon, involving many types of adaptations, but it is not as universal as parasitism. The parasitic relation may involve plant and plant, plant and animal, or animal and animal, as well as dubious organisms like viruses. Some members of almost all of the phyla of organisms have adopted the parasitic habit. Among animals the exceptions are easily listed: there are no parasitic sponges, few coelenterates, no echinoderms, few molluscs, and no vertebrates except the pygmy males of certain fish that live attached to the females of their own kind.

The difficulty of framing a clear definition of parasitism is best illustrated by citing examples of some of the kinds of food relationships that have been developed among the insects. There are clear examples of parasitism, such as the larvae of botflies that live under the skin of mammals. The botfly larvae are completely dependent on the mammal host to which, under normal circumstances, they cause a minimum of injury (though a lot of discomfort). Then we have clear cases of predation, like the robber flies that perch on grass stems ready to pounce on any smaller insect that happens by, sucking out its juices and discarding the useless corpse.

But with the so-called parasitic hymenoptera, whether to call the food relations parasitic or predatory is not so clear. There are several large families of insects related to the wasps that have taken up a special sort of food habit. The adult lays its eggs in the larvae of other insects. The number of eggs is delicately adjusted to the size of the parasite and host, and the young parasitic larvae when they hatch follow instincts that lead them not to eat essential organs, to leave the host alive until the parasites are full grown. Then they polish off what remains of the host, with a nice calculation so that the final host tissue serves as the last meal of the parasite larvae. This invariable complete destruction of the host is typically predatory, possible only when potential hosts are numerous (as they are, of course, with insects). Yet the internal life of the parasite larvae, their specialization for a very restricted type of host, are phenomena of parasitism.

The thing becomes very complicated, since there are parasitic hymenoptera that live on the parasitic hymenoptera that live on the other insects, and so even to the third degree. And the parent insects may show wonderful instincts for locating the potential larval hosts in the most inaccessible situations, more comparable with the instinctive developments that we associate with predation than with the "degeneracy" of parasitism. William Morton Wheeler has suggested that this special sort of insect food habit be called "parasitoidism".

Then we have the wasps which specialize in catching certain types of prey—caterpillars, grasshoppers, spiders, depending on the species of wasp—and stinging the captive so that it is immobilized but not killed. An appropriate number of stunned victims is stored in a nest cell, an egg laid, and the whole thing sealed off. The wasp larva can then, at its leisure, devour these stored provisions. This, of course, is a special sort of a predatory relationship, but it has many of the characteristics of the parasitoid habits of the wasp relatives.

And what do we call adult mosquitoes? They are adapted to suck the blood of larger hosts, usually vertebrates, but they live a free life, and attack a series of hosts in succession, like the robber flies, only the mosquitoes don't kill their hosts. The mosquito relationship is not clearly a predatory one, but neither does it have the characteristics that we associate with parasitism. Sir John Arthur

Thomson considers that relationships like those of the mosquito and the flea are better classed as predatory than as parasitic, and I think I agree, though this is contrary to the usual practice.

Our problem really is with the human mind, which needs to deal with discrete categories, even when these have to be imposed on an essentially continuous series. The trouble comes when we mistake the nature of our categories. Parasitism and predation are useful concepts, but their utility is destroyed if we get involved in a Procrustean process of making every instance fit one or the other concept. Science sometimes achieves rigid definition that can be handled with precise mathematical logic; but often also it must deal with vague and essentially undefinable concepts, where logic becomes a handicap. I think an important part of the scientific method is the ability to work with indefinite and provisional concepts, to accept approximations which can be used until something better is available. To be scientific, in this sense, is to be unsure, to be indefinite where knowledge or the nature of the material does not warrant definiteness. So we shall proceed boldly to discuss parasitism even though we cannot enclose it in a net of logical definition.

The sorts of adaptations involved in the parasitic habit can perhaps best be illustrated by a series of examples from various plant and animal groups.

VIRUSES AS PARASITES

All known viruses are parasitic, some on plants and some on animals. The plant and animal viruses may be different sorts of organisms, lumped together merely because of their submicroscopic size and obligate parasitism. Some of both, however, share one important characteristic, insect transmission. There are a large number of plant viruses, many of them causing diseases recognizable by the splotchy discoloration of leaves, the mosaic diseases. Most of these viruses have no way of getting from one plant to another except through the agency of sap-sucking insects such as plant lice, leaf hoppers and thrips.

Some are transmitted "mechanically", that is through virus particles adhering to the insect mouthparts, whence they can infect the next plant bitten. But generally the virus goes through a complete cycle in the insect, being taken into the stomach with the plant juices, passing through the gut wall, and eventually reaching the salivary glands whence it may be injected into new plants. This takes several days, and in such cases new plants will not be infected until the virus has become established in the salivary juices; once this incubation period has passed the insect may remain infective for a long time, transmitting the virus to many plants.

These viruses are highly specific. That is, a given kind of virus such as "tobacco mosaic B" will only be able to infect a very few different kinds of plants, such as tobacco and its close relatives. The viruses may also have a very limited number of insect vectors, perhaps only one species of plant louse. This seems to me an extraordinary thing, that a parasitic organism should be restricted to a particular kind of plant and to a particular kind of insect, considering how utterly different these two things, insects and plants, are.

There is no evidence that the plant viruses actually grow in the insect tissue. In fact the experiments indicate that an insect sucks up a definite amount of virus, and that this virus passes through its body without undergoing any increase. The property that makes one insect a vector and another not a vector seems to concern the gut wall, since a non-vector insect may be able to transmit virus if the gut wall is pricked with a needle after the insect has taken an infective meal of plant sap. There is also a definite limit to the amount of virus in the salivary glands, and this virus supply is eventually exhausted after the insect has fed on many plants. It is remarkable enough, however, that the virus is able to live indefinitely in the insect tissues.

Viruses are also the causative agents of a great variety of animal diseases. Many of these, such as measles and smallpox, are transmitted directly from host to host through contact. But one group of these viruses, including yellow fever, dengue and the so-called "sleeping sickness" of the United States (the encephalitides) is transmitted by mosquitoes.

The mosquito-transmitted animal viruses are present in large quantities in the blood at some time during the infection of the vertebrate host, even though they may chiefly multiply in some other kind of tissue. The mosquito thus may take up a large number of virus particles along with a blood meal. The behavior of animal viruses in the insect vector differs from the behavior of plant viruses in that the virus actually becomes established, multiplies, in the insect. Once a mosquito becomes infected, it can transmit the virus endlessly to new hosts as long as it lives. There is a sharply defined incubation period, as with the plant viruses, between the infectious meal and the appearance of virus in the salivary glands.

The animal viruses, like the plant viruses, are restricted to definite animal hosts and insect vectors. A given virus, like that of yellow fever, may have several different mammal hosts, and several different vectors, but without apparent properties in common, so that the only way to find out which mammals and mosquitoes are susceptible is to try the necessary experiments.

Curiously neither the plant nor animal viruses seem to cause any ill effect to their insect vectors, though other kinds of viruses may be responsible for disease conditions in insects.

BACTERIA AS PARASITES

Parasitism has been more thoroughly studied in bacteria than in any other group of organisms. The word series, bacteria—parasite—disease, is almost automatically linked. Yet the parasitic bacteria form a small minority of the group. The commonest food habit among bacteria is saprophytism, living on dead material; and in the bacteria the parasitic habit seems to have developed directly from the saprophytic habit, the transition being from life on dead meat to life on live meat.

Relatively few bacteria are obligate parasites, incapable of living apart from their hosts. This is very helpful in studying them, because even the disease agents can be grown in culture media in the laboratory, in gelatin provided with appropriate nutrient materials. A large number of different kinds of disease-causing bacteria will grow quite happily on slices of potato—one of the first culture methods developed by the great German bacteriologist, Robert Koch. Obligate parasites, like the viruses, many protozoa, parasitic worms and so forth are much more difficult to handle, and in many cases it has been impossible to devise methods of keeping them alive apart from their hosts.

Bacteria are responsible for a great many human diseases—such as typhoid, most types of pneumonia, diphtheria and tuberculosis. They are also involved in all of the pus-forming infections with which we are so familiar.

Disease, of course, is any departure from the normal condition (health) in an organism. Parasitism may occur without disease, as in the perfectly healthy person who is a typhoid carrier. Disease may occur without parasitism, as in the deficiency diseases like rickets, or in mental diseases. Or, as is most common, the particular symptoms that we recognize as a particular kind of disease may be the direct or indirect result of parasitism by a specific parasitic organism.

The disease, with bacteria, is often a very indirect result of the parasitism, caused by the toxins produced by the bacteria, rather than by direct damage from the feeding of the bacteria. Thus the organism that causes tetanus (Clostridium tetani) is not a parasite at all, since it cannot invade living tissue, but can only grow on dead tissue in deep wounds. The disease, tetanus, is caused by the effect of the toxin produced by this organism growing in such a situation. The toxin is one of the most powerful poisons known—0.00025 gram being sufficient to kill a man. Food poisoning (botulism) is likewise caused by a toxin rather than by the bacterium itself; and the disease diphtheria is due to a toxin rather than to the parasitic effect of the organism, although in this case the causative organism (Corynebacterium diphtheriae) does live a parasitic existence on the mucous membrane of the nose and throat.

Any discussion of bacterial parasitism leads very soon to a consideration of the mechanisms of defense on the part of the host, since the course of a bacterial infection depends so clearly on the interaction of the invader and the defense mechanisms. It is with bacterial infection that we can see most clearly the role of the leucocytes in destroying invaders, and the effect of antitoxins and antibodies in immobilizing the parasites and neutralizing their toxic products. But this leads us away from the level of natural history into problems that are more properly physiological.

FUNGI AND SEED PLANTS AS PARASITES

Parasitism by fungi is similar to parasitism by bacteria in that both groups are primarily saprophytic, living on dead organic material, and seem almost accidentally to get involved with parasitism by getting onto tissue that is still alive. The fungi are relatively unimportant parasites from the mammalian (human) point of view, but they are very important as parasites of cold-blooded animals and of higher plants.

Parasitism by seed plants—mistletoe and dodder—is of interest chiefly because it does occur, underlining the universality of the phenomenon. Thus the common dodder has lost chlorophyl, and has come to depend on its hosts for all food materials. Mistletoe has green leaves, and depends on its hosts chiefly for salts and water.

PARASITISM IN ANIMALS: THE PROTOZOA

It is in the animal kingdom that we find the fullest and most complex development of parasitism, the most ingenious systems for taking advantage of the energy of the other fellow.

Among the protozoa the simplest cases are perhaps those in which there seems to have been an easy transition from saprophytic to parasitic habit, as with the bacteria and fungi. Most amoebae for instance are free-living organisms, saprophytic, living on bits of organic matter that they can engulf, or perhaps predatory if engulfing smaller organisms like bacteria puts them in that class. Some amoebae have discovered the rich possibilities of life in the intestines of large animals, where they seem to cause no harm, living on the fecal debris. But a few have gone further, and learned to attack the tissues of the intestinal wall. One member of this minority, Entamoeba histolytica, has become the causative organism of a very uncomfortable and dangerous human disease, amoebic dysentery.

The parasitic amoebae differ hardly at all in structure from their free-living relatives. There is every gradation between such cases and the types of protozoa that have become greatly modified for the special conditions of parasitism. The malaria parasite, Plasmodium, whose life history was discussed in Chapter V as an example of the complications of reproduction, is a good example also of extreme specialization for parasitism among the protozoa. Plasmodium belongs to a class (the Sporozoa) the members of which are all parasitic. The class is divided into a series of orders and families which have become specialized for a wide range of parasitic situations in a variety of hosts, often with complicated life histories involving an alternation of hosts, as in the case of Plasmodium itself.

Parasitism has obviously had a long evolutionary history in this group, and I don't think its trend can be called toward "degeneracy." We habitually link parasitism and degeneracy, perhaps partly because of the human connotations of the terms, but partly also because with complex organisms parasitism is generally associated with a loss of sense organs and mobility which, necessary for the independent organism, become useless impedimenta for the parasite. Then too parasitism seems to be a road that cannot be retraced. Once a group of organisms has acquired specializations for living parasitically on other organisms, their fate is linked with that of the hosts.

But evolution in the Sporozoa has certainly not been toward degeneracy in the sense of simplification of life history or morphological form. An amoeba is a simple, straightforward animal as compared with a Plasmodium. The amoeba is either active, crawling around, or, badgered by an uncomfortable environment, it has pulled itself into a suspended cyst state, in which it passively waits for better times.

Compare this with the complicated and varied life of a Plasmodium. If we start tracing the Plasmodium life cycle in the vertebrate host, we find that it begins as a special stage (only recently discovered) in liver tissue whence forms are launched that are capable of invading the red blood cells. The red blood cell forms follow a developmental rhythm whereby they synchronously burst out from their old cells and attack new ones, giving the host his neatly periodic paroxysms. From time to time they give rise also to special sexual forms which stay in the blood cells unless, or until, they are sucked into a mosquito stomach, when they burst free to mate in this special gastric environment. The fertilized zygote then encysts on the stomach wall of the mosquito, produces a multitude of sporozoites which presently migrate to the salivary gland, whence they can be injected into some new vertebrate host, to take up the cycle again in its liver.

This can hardly be called degeneracy; but it is the result of a long line of evolution of the parasitic habit, an extreme of parasitic specialization. It seems also to have been "successful," insofar as there are a great many different kinds of Plasmodia, living in great numbers in a corresponding variety of hosts. The parasites and hosts for the most part have reached a balance whereby the parasite inflicts a minimum of injury on the host. No one has ever been able to demonstrate any deleterious effect from Plasmodium infection on the mosquito, measured by mortality or length of life of the mosquitoes, though one would think that the mosquito would at least get a severe stomach-ache from the process of encystment. Most Plasmodia seem also to cause a minimum of injury to their vertebrate hosts, lizards, birds and monkeys; though the four species of Plasmodia that parasitize man cause an appalling loss of human energy.

Parasitism occurs in several other groups of protozoa, besides the amoebae and sporozoa. Trypanosomes, a group of flagellate protozoa, are parasitic in a wide variety of animals and plants. An organism of this group is the cause of the notorious African disease, sleeping sickness. The trypanosomes of sleeping sickness are transmitted by a peculiar insect, the tsetse fly (Glossina). The larva of this fly grows to maturity within the abdomen of the parent. The fully developed larva is deposited on the ground, where it pupates. These flies thus must have about the lowest reproductive rate, and the most complete protection of the young, of any insect.

The spirochaetes are a distinctive group of organisms sometimes classed as protozoa, sometimes as bacteria. They all look like microscopic twisted threads, and their ability to move is the chief argument for including them in the protozoa. Many are free living in water. One genus is found in the intestinal tract of oysters and other molluscs, apparently harmless, and perhaps playing some symbiotic role. Other spirochaetes are parasitic, and include the causative organisms of several human diseases, especially syphilis and relapsing fever. The spirochaete of relapsing fever is transmitted by a particular genus of ticks (Ornithodorus) in some regions, by lice in others.

The parasitic worms (helminths) include several distinct classes of the phyla of flatworms and nematodes. The only parasitic annelid worms are the leeches.

The parasitic flatworms and nematodes show a great diversity of habit, and many have fantastically complicated life histories. An alternation of hosts is the rule: early development takes place in an intermediate host, and the sexual stage is reached in a definitive host. Thus the tapeworms (cestodes) commonly have some herbivorous animal as an intermediate host, encysting in the muscles of this host. Some carnivorous animal that eats the intermediate host gets the cysts along with the meat, and serves as a definitive host for the tapeworms that develop in his intestines. One of the commonest human tapeworms (Taenia solium) has the hog as an intermediate host—one reason for not eating insufficiently cooked pork.

The trematodes, or flukes, show the extreme of complicated life histories. The liver fluke of sheep (Fasciola hepatica) is the best known species. The adult form inhabits the bile ducts of sheep and goats (and occasionally of other animals, including man). These produce large numbers of eggs which are passed out with the feces and which, if they land in water, hatch after a week or two into an active, ciliated larva called a miracidium. This larva is capable of free life for a few hours, but there is no further development unless it finds a suitable snail host (some snail of the genus Limnaea). It bores into the snail where i: forms a sporocyst, a spherical sac-like body, which grows through the multiplication of enclosed germinal balls. These germinal balls form the next larval stage, the redia; when they are fully developed, the rediae burst the sporocyst and become free in the digestive gland of the snail, causing considerable damage to the snail. Each redia in turn finally forms several larvae of another type, the cercaria, which crawl out of the snail into the surrounding water. After a few hours, the cercaria wriggles a short way above the water on the stem of some plant, where it forms a hard cyst. These cysts, in a moist environment, remain alive for many months, undergoing no further development until they are ingested by some animal eating the plants on which they have been formed.

When eaten by a grazing mammal, such as a sheep, the cysts pass through the stomach unaffected by the gastric juices, and are digested in the intestine, freeing the next stage, called the metacercaria. These penetrate the wall of the intestine into the body cavity, and in about 48 hours start boring into the liver. They work their way slowly through the liver tissues, finally appearing in the bile ducts after about 7 weeks, ready to start laying eggs, thus completing the cycle.

This succession of stages outlines the general pattern of development in the trematodes, though there are endless variations among the 3,000 or so species that have been described. All of the digenetic trematodes pass through a stage in snail hosts, but that is about the only common denominator. There may be only one intermediate host (as in the case of Fasciola) or there may be two. In the latter case, both may be molluscs; one may be a mollusc, the other a crustacean; or the second intermediate host may be a larval insect; or a fish; or an amphibian. These occur in various combinations with definitive hosts—fish, amphibia, birds, mammals.

It is interesting to notice the various methods of getting from the snail to the next host. The cercariae, on leaving the snail, may penetrate directly in the definitive host, instead of encysting (this happens with the human disease called schistosomiasis). They may encyst on herbage, as in the case of Fasciola. They may encyst on the shell of the snail. They may encyst in the tissues of the snail, not undergoing further development until the snail is eaten by the definitive host. Or they may encyst in the snail, the cysts being deposited on herbage in "slime balls" from the snails.

One wonders that any trematode is able to meet the hazards of such a complicated life history, of finding each appropriate host in turn. It is not surprising to find that they have very high reproductive rates, to offset these hazards. The adult flukes produce large numbers of eggs (one species is said to lay as many as 25,000 eggs per day). But this is only the beginning, since there is an enormous multiplication within the snail. A single miracidium, becoming established in a snail, may give rise to 10,000 or more cercariae. Meyerhof and Rothschild kept track of the cercariae produced by a single snail infected with a trematode called Cryptocotyle lingua, and found that at the end of a year it had produced 1,300,000 cercariae. They kept this single snail alive for five years, and it continued during all this time to produce cercariae, though the rate fell off in later years to 800 or so per day. These reproductive rates differ greatly in different species of trematodes, of course, depending on the sort of hazards to which the particular species is exposed.

Some parasitic worms have simple, straightforward life histories. Thus the nematode Ascaris, which lives in the intestine of various mammals (including man), simply produces enormous numbers of eggs (said to be as many as 200,000 a day), which are passed out with the feces, on the chance that some of these will somehow get ingested again by some other suitable host. Even Ascaris, though, doesn't simply proceed to develop in the intestine of the new host. The larvae that hatch from the eggs penetrate the intestinal wall, and make a grand circuit of the body (blood, lungs, bronchi, trachea, glottis, pharynx) before coming back to the intestine to take up adult life. The various Ascaris found in different vertebrates look very much alike, but they are physiologically adapted to particular hosts. In an abnormal host, the larvae are destroyed in the course of their migratory circuit through the body.

ARTHROPODS AS PARASITES

Among the arthropods (crustacea, ticks, insects) parasitism takes diverse forms. Sometimes the adult is a parasite, like the curious barnacle Sacculina, parasitic on crabs. The adult Sacculina has become a mere sack of reproductive organs, absorbing nourishment from the crab through ramifying "roots" that penetrate everywhere in the body of the crab. This Sacculina, in its specialization for parasitism, has lost all of the distinctive characteristics that might indicate its place in the animal kingdom, and its relationship to the barnacles is deduced from the form of the early free swimming larva, which is like that of non-parasitic barnacle larvae. More frequently the arthropod larva is a parasite, as in the case of the botfly, while the adult form leads a free life. In some cases, as with ticks, the arthropod may be parasitic all through its life cycle.

CHARACTERISTICS OF PARASITISM

I have already made several remarks about the connection between parasitism and degeneracy. I suspect this is a matter of point of view. We are predatory animals ourselves, and consequently admire the characteristics of predation—agility, speed, cunning, self-reliance. We feel a certain kinship with the lion, and regard the liver fluke with horror. If a sheep were given the choice, though, it might prefer to be debilitated by liver flukes rather than killed by a lion.

The degeneracy of parasitism is specialization, involving the loss of structures and functions that are no longer needed. This seems to be, in evolutionary history, a one-way road, and the tendency of parasitism is toward ever greater dependency on the host. But this one-way development is a characteristic not only of parasitism, but of specialization in general. Adaptation to very special circumstances seems never (or hardly ever) to be reversible, and the reservoir of material for evolution in new directions lies among the proletariat of organisms with relatively generalized habits.

Parasitism may be a one-way road, but it leads into a vast territory. The thousands of species of groups like the Sporozoa among the protozoa, and the trematodes among the flatworms, probably represent single evolutionary stems which, through geological time, have become diversified into the distinctive orders, families and genera that we recognize today.

The parasitic habit has probably arisen in various ways. In many groups we can see an easy transition between the saprophytic and parasitic habits—between living on dead organisms and living on live ones. In other cases the transition may have been from some sort of predation. It has also been suggested that the shelter-seeking habit may lead to parasitism. Parasitism may also have arisen from symbiosis or commensalism, though here it seems to me that the path may have been two-way. It is just as easy to imagine symbiosis to have originated by the host of a parasite coming to depend on the parasite for some vital function as it is to imagine the parasitic relation to have arisen from a loss of reciprocity by one of the partners in a symbiosis.

Certainly the distinction between symbiosis and parasitism is often not clear. Many organisms that we are reasonably sure are parasitic, that is, living entirely at the expense of the host without direct return benefit, seem also to be entirely harmless. The distinction often depends on interpretation of adaptations. In the case of a micro-organism associated with a large organism, if the larger (the host) shows adaptations of structure or habit specifically favoring the micro-organism (like the mycetomes of insects), one assumes that the association is symbiotic. But if the host adaptations seem to be toward protection from the micro-organisms, toward the development of defense mechanisms, one assumes that the association is parasitic.

It is often said that the ideal parasite causes a minimum of injury to the host, and it seems obvious that it would be to the parasite's advantage to keep the host as happy as possible. A corollary of this is that where a parasite causes great injury to a host, the association must be a new one; that where the injury is slight, the association is an old one.

The general statement and the corollaries are probably both oversimplifications. Some very specialized parasites cause great injury to their hosts, others little or none; and equally, casual or accidental parasites may cause great injury or little. If one succeeds in the laboratory in establishing a parasite in a new kind of host, the parasite may cause an acute disease in the new host or the symptoms of attack may be subclinical—there is no way of predicting until the particular experiment has been tried. Similarly if a parasite like a virus or some bacterial strain is passaged repeatedly through different individuals of a new host, virulence may increase, may decline, or may remain at about the same level.

A devastating epidemic is probably always the result of a new or unusual set of circumstances, since violent and sudden shifts in population relations, such as result from a wildfire epidemic of a fatal disease, are not usual in nature. Such epidemics in man have been caused by the introduction of new parasites, or by alteration of environmental conditions surrounding old parasites. But man has altered natural population relations so violently in any case that it is hazardous to base any very sweeping generalizations on human host-parasite relations. There is no very objective evidence that man's association with the virus of chicken pox is any more ancient than his association with the virus of smallpox, though the two differ greatly in virulence.

A highly fatal disease caused by a host-specific organism must be adjusted in some way to the reproductive rate of the host, of course, or else both parasite and host would soon be eliminated. But adjustment of reproductive rates is probably at least as common an adaptation as loss of virulence. The fact that a lion kills a gazelle doesn't mean that the association between lions and gazelles is new; though the fact that both lions and gazelles exist together means that the two populations are in some rough numerical adjustment. The same, I should think, would apply to parasitic relations.

Host specificity; host response to invasion by a parasite; adaptations of parasites to the particular circumstances of host tissues and habits—the topics that could be discussed in connection with parasitism are endless. "Parasitology" is, in fact, a highly developed science in itself, which we could hardly subordinate to natural history if we tried. The prime task of the naturalist is probably to keep parasitism in perspective, as merely one example of the multitudinous relations among organisms. But since the naturalist, like all of the rest of nature, is commonly subject to parasitism, he is perhaps apt to view it as a calamity rather than as a matter of population adjustment, a method of getting the maximum diversity of kinds of life packets from the available raw material of the universe.

CHAPTER XI

The Behavior of Individuals

THE consideration of parasitism, predation, symbiosis and other intra-communal relations among organisms leads naturally to a consideration of the dynamics of population relations, to a review of the behavior of populations. But before taking up this subject, I think it would be well to insert a chapter on the behavior of individuals in which certain general topics useful in the understanding of population behavior can be dealt with.

To separate off chapters on behavior in a book on natural history is a rather arbitrary procedure, since one might reasonably argue that all of natural history is concerned with behavior. The word, according to Webster, covers "activity or change in relation to the environment"—which is pretty inclusive. Even our daily usage of the word is broad: we speak of the behavior of a ship in a storm, of the behavior of Japan in attacking the United States, of the behavior of electrons in an atom, of the behavior of little Gertrude who lives next door.

We are thus easily accustomed to think about the behavior of individuals, of groups, of things; about what they do and how they act. The prime interest of the naturalist in organisms is in what they do, how they behave, whether in growing and reproducing, or in relation to the physical environment, or in relation to each other as parasites, symbionts or parts of communities. In the last chapter we were concerned with the behavior of parasites, before that with the behavior of partners, before that with the behavior of members of communities. What, then, is left?

Actually, this method of treatment neglects a number of things that don't fit very readily into these particular pigeonholes, but that certainly must be included somewhere. It might have been more logical to have started the book with a consideration of the factors governing behavior in individual organisms, then to have taken up the behavior of organisms as parts of populations and communities, and lastly to have analyzed specific kinds of behavior like parasitism or symbiosis. But I doubt whether that sort of arrangement would have been any clearer. As things stand now, we are left with a group of behavior topics—territory, sexual behavior, social behavior—that perhaps gain a certain perspective if they are grouped together as aspects of the behavior of the individual organism.

TERRITORY

The robin bursting with song in the garden is not trying to impress his mate, nor is he the victim of mere exuberance, finding an outlet for accumulated *joie de vivre*. He is, rather, proclaiming the ownership of a territory, warning all stray robins that this area has been pre-empted and that no poaching will be allowed. This interpretation of bird song is now almost universally accepted; but the universal acceptance is rather recent.

To be sure, many of the early naturalists noticed that birds tended to have definite territories, but they failed to realize the significance of their observation. Margaret Nice, in a survey of the history of the concept of territory, found that John Ray, in 1678, reported that "it is proper to [the nightingale] at his first coming to occupy or seize upon one place as its Freehold, into which it will not admit any other Nightingale but its mate." Oliver Goldsmith, in 1774, actually used the word "territory", remarking that "the fact is, all these small birds mark out a territory to themselves, which they will permit none of their own species to remain in; they guard their dominions with the most watchful resentment; and we seldom find two male tenants in the same hedge together." A German ornithologist, Bernard Altum, developed the theory of territory in birds at some length in a book first published in 1868, but little attention was paid to his ideas outside of his own country.

The history of science is full of illustrations of the futility of being ahead of the times. The general acceptance of an idea depends on a receptive mental environment among the scientific community, on the ripeness of the time. Thus the general acceptance of the concept of territory for birds, mammals and other vertebrates dates from the publication, in 1920, of a little book by Eliot Howard, called *Territory in Bird Life*. Probably conditions for the acceptance of the concept had only then become ripe. Or perhaps Howard's influence was due to the beautiful exposition of his book—it is a fine example of the best natural-history writing. Whatever the explanation, the dating is clear.

The implications of the territorial habit are considerable. It means that the density of population of a given species of territorial bird or mammal (which means most birds and mammals) is determined not so much by food supply as by the availability of territory. The owner of a territory defends it against all intruders of his own kind and the surplus population, the individuals that do not own territories, are left at great disadvantage.

Margaret Nice recognizes a variety of classes of territory among birds, which she lists by the alphabet from A to F. In the first (and commonest) class the territory covers the mating, nesting and feeding ground. Class B includes mating and nesting areas, but not the whole feeding ground. Class C forms a mating station only. In Class D, the territory is limited to the immediate surrounding of the nest, and has different characteristics in colonial and solitary species. E includes winter territories, and F roosting territories.

The meaning of the territorial habit is far from clear. As Margaret Nice remarks, "it is based primarily on a positive reaction to a particular place and a negative reaction to other individuals." It seems at first glance to be a sexual phenomenon, particularly in birds; but on the other hand, many birds show territorial habits when they are not in a sexual phase, and in both birds and mammals, territory may be a possession of a flock or band, including various males.

The effect of territory, certainly, is to space the population of a given species, putting a fairly definite upper limit on its density in a given area. This results in assurance that, under normal circumstances, the food supply will be adequate. Whether territories are generally in excess of what would be needed to provide food is a matter of debate, and probably varies with different species.

It seems likely that Mrs. Nice is right in maintaining that "the chief function of territory is defense—defense of the individual, the pair, the nest and young." An animal on its home territory has a physical as well as a psychological advantage. It can know the terrain intimately, where to hide, where to find food. The territory owner, in defending this territory against intruders of his own kind, selfishly insures his own survival and dooms the ejected intruder: but in so doing, favors the survival of his species by tending to maintain an optimal density for that species.

Paul Errington, who has made detailed studies of animal population behavior in our Middle West, has summarized a mass of observation tending to show that it is largely the territory-less individuals that are subject to predation, and thus that population control depends not so directly on the activities of predators, as on the availability of territories. The territory owners are relatively immune to attack, but the excess, floating population is doomed, and the particular kind of predator that serves as the agency of doom is more or less incidental.

There is a great deal to be learned about territory in animals; it is one of the rapid growing points of natural history. Territorial habits of some sort are widespread among vertebrates—fish, reptiles, birds and mammals—but whether comparable habits are to be found among invertebrates is doubtful. Among vertebrates, the accumulated observations on sexual and social behavior would surely gain new perspective if re-examined with the concept of territory in mind.

SEXUAL BEHAVIOR

Active defense of individual territory may be limited to the vertebrates, but complicated sexual behavior is an almost universal phenomenon in the animal kingdom. Most of Darwin's famous book on the *Descent of Man* is taken up with a discussion of "secondary sexual characters"—characters of structure and characters of behavior—and this book remains an admirable summary of the observations. Much additional material has been accumulated since Darwin's day, but the new material differs little in kind.

The deductions that Darwin drew from this material—forming the theory of sexual selection—have received scant support from twentieth-century authors. Modern students of evolution have mostly been concerned with the mechanism of heredity, and they have generally studied laboratory animals like mice and the fruit fly, Drosophila, in which secondary sexual characters are not conspicuous. In birds, the group from which Darwin drew much of his material, we now know that song, conspicuous colors and aggressive attitudes are directed toward the defense of the territory from other males, rather than toward the winning of the female, and this has given strong support to a general tendency to look for nonsexual explanations of such "sexual" characters.

But even when we discount all of the effects that overenthusiastic naturalists have attributed to sex, secondary sexual characters of structure and behavior still remain a remarkable and widespread phenomenon, surely with some intimate relationship to the general problem of evolution. By secondary sexual structural characters, I mean all of the striking structures that are restricted to one sex, but that are not directly involved in the transfer of gamete materials in fertilization whether by copulation or some other method. The difference between the peacock and the peahen, the bright colors of many male butterflies, the scent glands confined to one sex, the horns of rhinoceros beetles or the antlers of deer are examples.

By secondary sexual behavior, I mean all of the peculiar antics that animals go through before they get down to the business of copulation itself: the drumming of grouse, the dance of a male manakin on his court, the swarming of male mosquitoes. And, I suppose, the fertility dances of Hottentots (if they don't have them, plenty of primitive people do) which perhaps persist into civilized society in a relict form as debutante balls.

Probably all of these things are mechanisms to insure that mating will occur within the species, by arranging that mating will occur under special circumstances in which the two sexes of a given kind of animal are brought together. At least it is certain that sexual behavior and secondary sexual structures may differ sharply in species of animals that are otherwise very closely similar.

This has been brought home to me by my work with mosquitoes. In general, these insects will mate only when the males are in a sexually excited state in which they form compact flying swarms. Each species of mosquito requires a particular set of circumstances for the formation of such swarms, especially a preferred light intensity, and a particular type of spatial orientation. Thus some swarm in daylight, others early or late in the evening. Some swarm low, over grass; some high in trees; some over moving objects such as large animals. In the laboratory, these mosquitoes will not mate unless and until the special environmental circumstances required for swarm formation are provided. A few species are not exacting in their requirements, and the males will start sexual flights in almost any kind of cage; many kinds will not mate under any circumstances that have so far been devised in the laboratory; still others will mate only with special arrangements of lights, or when provided with some contrasting object such as a mirror or a sheet of paper over which the swarm will be oriented.

The point I want to make is that each different kind of mosquito has different mating habits. The fact that you can arrange circumstances under which Anopheles labranchiae will mate in the laboratory is of little help in devising circumstances for Anopheles messeae, even though these two mosquitoes are so much alike in appearance that they can be differentiated only by special techniques (in this case, microscopic examination of the eggs!). The divergence in sexual behavior here is basic to the split of these mosquitoes into two independent populations, two species, that inhabit the same areas. If, by some trick, the student manages to induce labranchiae males to mate with messeae females, the resulting hybrid is sterile, like a mule. It is thus obviously important, for the maintenance of the species, that the males mate with the right kind of females. Hence, perhaps, the restriction of mating in each case to special and characteristic circumstances.

It is interesting that even in birds, sex recognition and mating within a species depend on the coincidence of behavior patterns in the two sexes. Many kinds of male birds, in breeding condition, will attempt to mate with dead, stuffed females. This provides many possibilities for experiment. Professor J. A. Allen of Cornell discovered that if he presented a male ruffed grouse with the skin of another male, laid on the ground, the prostrate condition of the dead grouse was "female behavior" to the live male, which attempted copulation. Similar experiments with the house wren showed that the male could not distinguish between stuffed males and females of its own species, or between its own females and those of closely related species, such as the winter wren. A stuffed specimen of a very different species, such as the marsh wren, was, however, given no attention.

It is interesting that even in birds, sex recognition and mating are just the first steps in an elaborate sequence of social behavior. Sexual behavior, where it involves contact between two individuals of the same species, might, in fact, be regarded as the simplest and most universal type of social behavior, the basis of all of the other complex relations that may be superimposed. The next step would be parental care, and when the sexual and parental relations are continued beyond the immediate necessity for the protection of the young, we have social behavior in the strict sense.

Aggregations of individuals of the same species occur commonly in all groups of organisms—a colony of corals, a forest of one kind of tree, a swarm of locusts—and the dividing line between such aggregations and animal societies is perhaps arbitrary. The forest of a single kind of tree and the reef of a single kind of coral are aggregations based on mere propinquity, and the relations between individuals of the aggregation are of the sort that would automatically result from this spacing. The swarm of locusts represents a transient and rather fortuitous aggregation. The genesis of a locust swarm, and the factors governing the behavior of a swarm once established, provide a fascinating and by no means simple field of study: but it is not considered to be study of social behavior.

The boundary between a mere aggregation and a society of organisms is generally considered to lie somewhere between a swarm of locusts and a school of fish or a flock of parrots. At least students of the latter sort of groupings of vertebrates are apt to consider that they are involved in the study of social behavior.

Whatever the beginning, there is no question about the extreme of social development: it occurs in the termites and ants; and in man. The importance of social behavior in man gives the whole subject a special interest.

I have always been very fond of a theory developed by Carveth Read, to which no one seems to have paid much attention, to the effect that man in his recent evolution must have gone through a hunting-pack stage, more similar to wolf packs than to the tribal and family groupings of monkeys and apes. This, according to Read, explains the remarkable similarity between men and dogs in behavior and emotional expression. He aptly quotes Galton, who remarked that "every whine or bark of the dog, each of its fawning, savage or timorous movements is the exact counterpart of what would have been the man's behavior, had he felt similar emotions. As the man understands the thoughts of the dog, so the dog understands the thoughts of the man, by attending to his natural voice, his countenance, his actions." Which reminds me of the theory of someone to the effect that the man-dog association was started by some clever wild dog adopting man, rather than the generally assumed reverse.

In the United States a group of scientists associated with Professor R. M. Yerkes of Yale has given especial attention to the study of social behavior in vertebrates, working mostly with anthropoid apes and other primates. These Yale studies have been well publicized by the inimitable Earnest Hooton (of Harvard). Another center for the study of social behavior has formed at Chicago, around Professor W. C. Allee, who has written several interesting books on the results of his studies. Allee, among other things, has given considerable attention to the study of the "peck order."

THE PECK-ORDER

The "peck-order" was first clearly described by a Norwegian named Schjelderup-Ebbe who wrote a paper in 1922 on his observations of the social hierarchy in a flock of chickens. The similarities between the behavior of chickens and humans must have impressed many a farm boy—the conversational cluckings, the bluff and posing of the big shots, the urgent rush in time of stress to get to the other side of the road, and the aggressive dominance of certain self-confident birds who "rule the roost." This last phenomenon, following the lead of Schjelderup-Ebbe, has been studied by a whole flock of animal behaviorists.

Who pecks whom, in a chicken run, turns out to depend on a fixed hierarchy which is determined early in the history of a particular flock. As Allee says, "When two chickens meet for the first time there is either a fight or one gives way without fighting. If one of the two is immature while the other is fully developed, the older bird usually dominates. Thereafter when these two meet the one which has acquired the peck-right, that is, the right to peck another without being pecked in return, exercises it except in the event of a successful revolt which, with chickens, rarely occurs."

There is usually one bird which can peck all of the others, another which can peck all except the first, and so down the line to the last poor devil who is pecked by everyone. Sometimes, though, triangular situations develop, in which A pecks B, B pecks C, but C pecks A. If a bird is removed from a flock for a long time, and is then returned, its status is determined anew, sometimes with disastrous results for the returning bird.

Schjelderup-Ebbe extended his study to many other types of birds, and found in all cases such a peck-order among individuals. He thus became convinced that despotism was a major biological principle. "Despotism," he stated, "is the basic idea of the world, indissolubly bound up with all life and existence. On it rests the meaning of the struggle for existence."

The despotic nature of the peck-order among chickens has been amply confirmed by all who have studied the subject, but few biologists would agree that this makes despotism a general biological principle. Allee has studied many other species of birds, and, unlike Schjelderup-Ebbe, he has found the fixed hierarchy of chickens to be the exception rather than the rule. In birds with strong territorial habits, an individual may be despotic in his own territory, though a Milquetoast when he has passed the boundary. And in other flocking birds, such as pigeons and parakeets, the peck-order may be far from clear and apparently subject to frequent variation.

Allee and others have made some especially interesting observations on "leadership" among these flocking birds. The characteristics of leadership, as in a formation of flying birds, are by no means easy to discover, but there is at least considerable evidence that the bird in front, the apparent leader, may not determine the maneuvers of the flock. Change in flight direction, for instance, may be initiated at some other point in the formation, the "leader" merely, in this sense, following the direction taken by the mass behind him. This, of course, hardly detracts from our admiration of the beautiful precision with which some kinds of birds fly and wheel in formation—a type of social behavior that has hardly been touched as a field of study.

SOCIAL LIFE AMONG THE INSECTS

Social life among the insects is on a different plane from that among the vertebrates, out of this world, serving H. G. Wells aptly for schemes of life on the moon or on Mars. The individual is completely subordinated to the community which is really a vast family with one mother, the queen; a community that acts as a superorganism, its members rigidly, blindly, instinctively carrying out their appointed roles. An ant hill or a termite nest represents perfect social organization—which man, thank goodness, seems incapable of achieving. The social philosophers who consider the State to be the unit, the perfect functioning of the State the goal, should study the ants, who have long since solved this problem. Surely the genius of man is to find some social nexus in which the individual can maintain his sense of importance, because the consciousness of the individual seems to be the essential, unique human contribution to the universe: the individual with his hopes, his fears, his happiness, his preposterous self-consciousness so out of perspective with all of the rest of nature.

It is, of course, probable that the individual ant, if possessed of "self-consciousness" and "self-expression," would be convinced of his own individual importance in the functioning of the ant hill, and would deride our dismissal of him as a completely subordinated unit. Perhaps we too are caught in a development of evolution that has seized the social organization, rather than the individual, for its raw material, and realization of this might come as hard for us as for the ant.

But this is wandering afield from natural history—though I am not sure where the boundary line should be drawn, in behavior studies, between natural history and physiology and psychology. The three words—and the corresponding sciences—perhaps represent little more than somewhat different approaches to the same problem, that of how organisms behave, and why. With the social scientists the shift is more in subject (man) than in point of view.

TROPISMS

The naturalists (ecologists) have generally been most interested in field observation of behavior; while the physiologists and psychologists have been most interested in laboratory analysis of behavior in controlled situations. The complete understanding of the behavior of the organism will depend, of course, on a synthesis of both types of study, a synthesis that should surely be a function of the broadened naturalist, in his effort to build up a picture of the interactions of individuals and populations in communities.

Both the psychologists and physiologists have been much concerned with the reactions of animals and plants to various environmental stimuli. The study of such isolated reactions sounds simple, but it has grown into a very complex and specialized field of study, with one of the most forbidding vocabularies to be found among the biological sciences.

It started with de Candolle who, in 1832, coined the word "heliotropism" for the phenomenon of plants like the sunflower turning toward the sun. The botanists soon found that many types of plant behavior could be thus described in terms of simple responses to specific stimuli: that roots grow downwards in response to a gravitational stimulus, being positively geotropic; that in some cases roots will grow toward an area of soil with the greatest concentration of some mineral salt or other, showing positive chemotropism; that tendrils on a vine tend to twist in response to contact, showing thigmotropism; and so *ad infinitum*.

Jacques Loeb was more responsible than anyone else for the wide adoption of the vocabulary of tropisms in studies of animal behavior. Loeb had an almost passionate desire to show that all animal behavior could be interpreted in terms of simple forced responses to particular environmental stimuli. He summarized his ideas in a book published in 1918 called *Forced Movements, Tropisms and Animal Conduct,* which has had great influence, both by making converts and by making enemies.

Loeb's concept of the role of tropisms in animal behavior is surely an oversimplification. But I doubt whether things are helped much by the current tendency to compensate the simplification by a complication of vocabulary whereby tropisms are replaced by "kineses, taxes and transverse orientations." The purpose, of course, is to gain objectivity by avoiding words that have any connotations in terms of human conduct. The students of human behavior, presumably, want to find words that have no connotations at all.

Automatic and fixed responses to specific stimuli (tropisms) are certainly of basic importance in the behavior of invertebrates, and they make beautiful subjects for experimental study. The moth flying to the light is showing a tropism (I suppose a positive phototaxis, in the newer vocabulary), and a particular kind of moth in the same physiological state will always exhibit the same response to the same type of stimulus. Similar fixed responses can be shown in vertebrates, though with vertebrates such a response is more commonly called a "reflex." But in vertebrates this fixed behavior is extensively modified by memory and experience, by "conditioning."

With the words "conditioned reflex" we have clearly reached the territory of the psychologists—a dangerous land with vast marshes of philosophy and tangled jungles of controversy. It is a land that all thinking men should explore, because perhaps therein lies our salvation as a species; but my services there as a guide would have no value. We had better, in the present book, get back to regions that have been more definitely charted by the naturalists, going, with some logic, from the behavior of the individual to the behavior of the population, of the species.

The common denominator of the various kinds of behavior that have been mentioned in the present chapter is a method—that of watching organisms, recording what they do, and attempting to find explanations and general patterns. This common method holds together the physiologist measuring the response of a cockroach to a smell conveyed through complicated apparatus under rigidly controlled conditions; the psychologist studying the speed with which a mouse learns to finds its way to food by running through a maze of dead ends and false turnings to the rewarding goal; the ornithologist huddled in his blind, noting down each movement of the robin as it goes about the daily chores of living. These people are all watching individuals, and from the accumulations of individual observations, trying to build up a general picture for the species, for the process, or for the working of living things in general.

When we pass to the study of the behavior of populations, these objectives are hardly changed, but it is increasingly necessary to have recourse to an additional method, that of statistics, because the unit of study, the population, is an abstraction that cannot be caught, or watched from a blind.

CHAPTER XII

The Behavior of Populations

AT first sight it seems a little odd to write of the behavior of populations. A population is unlikely to be dancing a waltz or climbing a tree or to be found making love to another population. Yet neither is a population a fixed, static thing. Any population grows, develops relations with other populations and with the physical environment: it may subdivide into new populations, or, in the course of time, become extinct.

I don't know whether there is any point in trying to define a word like population. I am using it here for the aggregate of individuals of any one kind living at any one time—sometimes restricting it also to some particular area, sometimes using it for the aggregate of the same kind of individuals wherever found. This is a common usage. We can write that the human population of the earth is estimated at two billions; or we can write about the population of New York, or about the Negro population in South Carolina. We habitually indicate the kind restriction and the area restriction, and still find population a very useful concept.

The behavior of a population is not the sum nor even always the average of the behavior of the individuals making it up. In dealing with populations, we have changed the level of generalization. Just as the individual is something more than the sum of its cells, so is a population (or a community) something more than the sum of the individuals making it up.

In dealing with populations and communities we come at once in contact with the statistical method. And we have to cling to the statistical method, whether we like it or not, because it is our only life preserver in a dangerous sea of abstractions. An individual can be watched, his actions described, his development followed. The individual represents what Warren Weaver calls a "problem of simplicity"—though the problem may seem complex enough at times to the observer. But the population escapes direct observation as a whole; it must be followed by means of samples. The antics of hawk D or fox X by themselves are of no use to us; we have to find out what hawks or foxes do on the average, and find out the effect of this, on the average, on rabbits. We have to determine this from a sample. This may be a very hazardous business as the *Literary Digest* and later Mr. Gallup found out. We are dealing with problems of disorganized complexity and, worse yet, of organized complexity.

For present purposes I don't think we need to become involved with the mathematics of the problem. The most essential point is to realize that, in describing population behavior, we are dealing with dynamic, not static, phenomena. It is hard to remember this because phrases like "the balance of nature" come so easily and seem so plausible.

CYCLES AND PLAGUES

In one way, certainly, the biotic community seems to be in a state of balance. The numbers of foxes and the numbers of mice seem to be mutually adjusted. But if we take a census regularly of the foxes and mice in a given area, we find that the numbers of both show wide fluctuations from year to year. In many cases where data over long periods are available, these fluctuations are cyclic.

The English ecologist Charles Elton has devoted especial attention to this problem of fluctuation in animal numbers, going over old records wherever they are available, such as those on furs kept by the Hudson's Bay Company. These records show that there are peaks in snowshoe rabbits, for instance, about every eleven years, with corresponding peaks for lynx falling a year or two after the rabbit peaks. Other animals, such as the Norwegian lemmings and the arctic fox, may show peaks at shorter intervals (about three years apart in both these cases). One mammal on which records are available, the beaver, seems to be free of these periodic fluctuations in population density.

We really have no very clear understanding of these cycles of animal abundance. Attempts to correlate animal cycles with climatic cycles have not been very successful—but then, the whole subject of climatic cycles is not well understood.

Particular animal populations sometimes reach tremendous numbers and erupt as plagues. This is particularly true of mice and lemmings, whose fluctuations and migrations form the subject of a book by Elton. Locusts show the same sort of phenomenon. In such cases, the pressure of built-up numbers seems to result in an explosive migration into new areas, where the migrating individuals die. It is a sort of mass suicide. I have watched this with South American butterflies—standing fascinated while millions and millions of individuals streamed past me out to sea, to certain death. Of such is the "balance of nature."

POTENTIAL REPRODUCTION

The actual number of organisms of a given kind existing at any one time is the result of the interaction of the potential rate of increase of that species and the environmental resistance—the rate at which individuals are killed off. Now the potential rate of reproduction of any particular kind of organism is fairly constant, but the rate at which individuals are killed off may fluctuate greatly, which results in corresponding fluctuations in the numbers of individuals alive.

The reproductive rate of organisms varies tremendously. An elephant may produce one young every three years or so. A bacterium of a species that divided every hour would, if each individual lived, have produced 16,777,216 new individuals in 24 hours. A few examples quoted by Dobzhansky give an idea of the possible maximum reproduction of single individuals of many kinds of organisms: an individual of the fungus Lycoperdon bovista produces 7×10^{11} spores (seven hundred billion); a single tobacco plant, 360,000 seeds; a salmon, 28,000,000 eggs in a season; an American oyster up to 114,000,000 eggs in a single spawning.

Calculating the progeny of an individual under "ideal" conditions makes a nice game, and textbooks are full of the results: how long it would take the progeny of an individual fly to equal the weight of the earth; how many times the offspring of a single beetle in a single year if laid end to end would reach to the moon and back. With a pencil and paper, you can rapidly discover that the place, in no time at all, would be overrun even with elephants, if they all lived. Darwin made an elephant calculation. Assuming that breeding begins at the age of thirty and continues to ninety, and allowing six young for each female elephant in this period. Darwin figured that the progeny alive from a single pair, after 750 years, could be ninety million.

Potential reproduction follows an exponential curve—1, 2, 4, 8, 16—and exponential curves, by the nature of things, can only be maintained for brief periods. They run off the paper, or head into interstellar space. As Stuart Chase says, "it is a fearful thing to be aboard an exponential curve—something like an express train out of control." It must, in no time at all, hit something. In the case of population growth, it hits the environmental resistance. The worries of students of human population, of the prophets of gloom in the chain from Malthus to Vogt and Osborn, are based on the fact that, in man, we have been tampering with the environmental resistance—fooling with the brakes on our exponential express, without giving corresponding attention to methods of reducing the speed.

THE TEETERING BALANCE OF NATURE

The balance of nature, then, is an adjustment between the potential rate of reproduction and the environmental resistance. How steady the balance appears depends on how closely we look at it. If we sit comfortably on a hill and contemplate the forest below, it may seem very steady indeed. The elephant, safe from predators because of his size, group habits and effective tusks; protected from parasites by a complicated antibody system; freed from climatic hazards by adjustments of mammalian physiology like those controlling the temperature of his blood stream, meets a very low environmental resistance with a very low reproductive rate. And so all through the catalogue to the bacteria that are so constantly reproducing and so equally constantly dying from any of thousands of kinds of trivial accidents like getting stranded in a fleck of sunshine or ingested by a passing amoeba.

With the perspective of geological time, the balance would be seen to have a steady bias, weighted in one way or the other, as different kinds of populations built themselves up to numerical abundance or waned away into extinction. This might be related to the slow changes of environmental resistance brought about by the geological shifts of climate, perhaps directly, or through the changing biological relations that would necessarily ensue. There must also be slow changes in environmental resistance through the process of evolution itself, new types of organisms, new adaptations, new combinations giving advantages to different sets of populations.

We have then these great swings of the balance in geological perspective, contrasted with its apparent steadiness with the perspective of the man on the hill. But if we examine the balance with the reading glass of statistical counts of population density, of census methods, we find that it is a very jittery balance indeed, with seasonal swings, cyclical swings, and a lot of just plain random jerks.

These swings and jerks interest the ecologist, because through studying them he can learn a great deal about the environmental relations of organisms. Since the potential reproductive rate is, as I said, a steady affair within the available geological period of observation, the unease of the balance must be due to fluctuations in the environmental resistance, hard times and good times for whatever population is under the reading glass. The analysis of this situation thus presents all sorts of interesting possibilities.

THE DYNAMICS OF POPULATIONS

The time element makes studies of organisms like the larger mammals difficult: observation is almost necessarily limited to fur-bearing species on which statistics are available, like those collected by Elton from the Hudson's Bay Company. Students of wildlife management have, however, developed a keen interest in the subject, and long-term observations on many species of birds and mammals are now being carried out in various parts of the world.

The insects have received most attention from students of population behavior. With many generations a year, significant variations in population density can be followed within convenient time limits, on the basis of one or a few years of study. Insects are small enough, too, so that miniature populations can be established in the laboratory under experimental conditions, and various factors affecting the growth and decay of the populations studied. Fruit flies (Drosophila) and flour beetles (Tribolium) have been favorite subjects for such studies, and a surprisingly large amount has been written about the behavior of laboratory populations of these insects.

One objective has been to find methods of describing the

fluctuations of populations in mathematical terms. Science progresses, in large part, in so far as it succeeds in finding methods of measuring phenomena, and of finding relationships among the measured phenomena that can be expressed in the abstract terms of formulae. Mathematics is said to be the language of science, and the physicists and chemists have succeeded extraordinarily well in describing the events that they observe in this language. The events of biology have proved to be more refractory, but this hardly means that it will never be possible to describe them with mathematical symbols. Population problems seem, in fact, to offer one of the most attractive openings for the entry of mathematics into biology.

Elaborate studies have not been confined to the insects. One of the favorite subjects, the one certainly with the most massive accumulation of literature, is man himself. The study of the behavior of human populations forms in fact, a whole nicely demarcated science to itself, demography, with textbooks, journals, associations of devotees, and all of the rest of the paraphernalia of an independent scientific discipline.

The demographers have come to feel more closely associated with the social sciences than with the biological sciences; but man, inescapably, is an animal (whatever else he may be), and it is hardly practical to divorce the study of man from that of the rest of nature. Demography, actually, has many close connections with animal ecology. Raymond Pearl, one of the outstanding students of human biology, was equally at home with jars of fruit flies and with statistics from the Bureau of the Census, and he and his associates were all interested in arriving at generalizations about population phenomena that would be equally applicable to bacteria, Drosophila or man. This attitude is implicit in the title of a book that he wrote which deals almost exclusively with human problems, *The Natural History of Population*—a book, by the way, that serves nicely as an introduction to the general subject of population studies.

OVERCROWDING, UNDERCROWDING AND THE OPTIMUM

The behavior of populations, despite the mathematical aura that surrounds it, is hardly more exactly known than any of the rest of biology, nor are its concepts more precisely defined. What, for example, is the optimum density of population for a given kind of organism under given conditions? It seems likely that there is an optimum for each sort of situation. Unfavorable effects from overcrowding are easily demonstrable for many kinds of organisms, under both field and laboratory conditions. Overcrowding may be harmful not only because of the limits of the food supply, but because of many sorts of indirect effects—the creation of unusually favorable conditions for parasites or predators, interference with normal behavior patterns such as the spacing of nests, and so on.

Unfavorable effects from undercrowding are less obvious and less easily demonstrable, but none the less real. The extreme case, of course, would be a population so scattered that the two sexes failed to meet for reproduction, but Allee, who has given particular attention to the subject, has found a number of less obvious harmful results of undercrowding. Where there is a toxic or unfavorable substance in the environment, for instance, the effect may be more damaging if only a few individuals are present. Allee showed this under laboratory conditions in experiments with such things as goldfish in aquaria containing colloidal silver, water fleas in alkaline water, flatworms exposed to ultra-violet light. He also has been able to demonstrate, in a variety of organisms, a positive stimulus to growth if a few (but not too many) organisms of the same kind are kept together. Goldfish again grow better, in a normal medium, if several are together than if kept in isolation. A whole variety of effects can be shown in animals that ordinarily flock together, effects based on mutual stimulation and things of that sort.

Thus we can be pretty sure that there is a minimum density for organisms, below which unfavorable effects begin to be felt, undercrowding; and that there is a maximum density, above which the unfavorable effects of overcrowding begin to appear. Somewhere between would lie the optimum density, the spacing most favorable for the species. Obviously, though, it would not be easy in any given case to define this optimum precisely in terms of so many organisms per unit of area.

The difficulties of the concept of optimum in general are brought home if we try to think about the optimum density for human populations. Almost everyone would agree that some areas are overcrowded, and from many points of view a good case could be made out for some areas as being undercrowded. But what is the optimum? The manufacturer, interested in cheap labor, would have one idea; the health official would have another; the educator, faced with the problem of reaching a sparse population, yet deploring the effect of urban concentration, might differ from either; the wildlife people, who have the reasonable argument that it is valuable to have room left for other kinds of organisms besides man, would think still differently. The optima from all of these points of view would fall somewhere between the thin nomadic density at which social forces are hardly operative, and the density of clear overcrowding, at which natural resources and food supply are endangered. In other organisms besides man, the definition of the optimum would probably vary within similar limits, depending on the type of activity of the organism under consideration.

THE NUMBERS OF INDIVIDUALS

Population densities under natural conditions vary enormously with different types of organisms, and among different kinds of the same type. The number of individuals making up the species population of a kind of bacterium or alga, might be a very different sort of thing from the number of individuals making up a population of insects, which in turn might be vastly greater than the individuals in a population of mammals or birds. Mostly we have no very precise figures, because of the difficulties of making a census. We know the figures for man better than for any other species—there were estimated to be about two billion human individuals living on this planet in 1940. Probably no other mammal approaches this total figure, for while some of the small rodents may teem in their native haunts, none has the tremendous geographic spread of man. David Davis, who has been studying the population relations of rats in the city of Baltimore, tells me that this city, with a million humans, has only between one and two hundred thousand domestic rats. Thus even these ubiquitous creatures don't keep up with man—though of course Baltimore may have less rats than some places.

Many mammal populations are made up of a definitely small number of individuals—the orangoutang, the giant panda, the mountain gorilla, to cite obvious examples. The American bison, according to the calculations of Martin Garretson, reached a low in 1889 of 1,091 individuals which, by 1930, had built up to well over 21,000 (mostly in Canada). The population before the advent of European man on this continent was surely somewhere in the millions.

Birds have been much subject to censuses because of the interest of enthusiastic amateur bird watchers, and in consequence a great deal of information has accumulated on the fluctuations of the populations of various species from year to year. The accumulated studies up to 1944 were summarized in a paper by Kendeigh. Again most of the figures are relative, this year compared with that, or this place with that place, because of the almost insurmountable difficulties of arriving at an estimate of the total number of individuals for any given species at any given time. One of the few attempts at estimating total populations was made by Forbes and Gross for the state of Illinois. They arrived at an estimate, under midsummer conditions, of 30,750,000 native birds of all species, and of 5,536,000 English sparrows. The human population of this state according to the 1940 census was 7,897,241 individuals, somewhat in excess of the English sparrows.

Man has had a very good opportunity to watch the process of the extinction of populations of various birds and mammals; but though he has contributed so much to the process, the data available are rather meager. It seems likely that in many cases species have become extinct not so much because man as a predator has killed them off, as because he has altered the biological equilibrium through his alterations of the landscape. This was probably a factor in the famous case of the passenger pigeon, more surely a factor in the case of the ivory-billed woodpecker, as the Harvard ornithologist, Ludlow Griscom, has pointed out.

In the case of the heath hen, which finally died out under full ornithological supervision on Martha's Vineyard, we have a nice documentation of the process, which has been summarized by Allee in connection with his studies of undercrowding. In 1907, when sustained efforts to avert extinction were started, the population had been reduced to 77 individuals; it built up to 2,000 or so by 1916, when a combination of fire, gales, a hard winter, and an invasion of hawks reduced the numbers to fifty pairs or so. There were ups and downs for a while—reaching a high of 314 in 1920, then following a fairly steady decline year by year—117, 100, 28, 54, 25, 35, 20, until 1928, when only one bird was found, the last to be seen.

ADJUSTMENTS TO THE SEASONS

The adaptations of organisms to the annual climatic cycle can perhaps well be discussed as an aspect of population behavior. In organisms with a long life span, such as trees, birds and mammals, these adaptations are matters of individual behavior rather than population behavior—dormancy, migration, hibernation. But in organisms with a short life span, such as insects, adaptations to the changing seasons involve major cycles in population density as well as individual changes in habit, and are clearly a matter of primary concern in the understanding of the dynamics of the populations of these organisms.

Few habitats have a stable climate through the year. The depths of the oceans, tropical seas and some tropical forests and lakes, come closest to complete stability. For the most part, even in the tropics, the year is divided into well-marked seasons, sometimes an alternation of single dry and wet seasons; sometimes (close to the equator) the year is divided into two wet seasons separated by short dry periods. These dry seasons in many regions are very severe: forest trees shed their leaves; streams, marshes and ponds dry up; temperature and humidity relations are greatly altered; light intensities are changed by general haze.

But the most striking seasonal changes of the tropics are mild affairs compared with the changes of the temperate zone with winter, when temperatures fall below the freezing point of water. Under these conditions, all organisms must either have some method of maintaining internal temperatures above the freezing point; or must enter some special state in which the life process can be maintained, dormant, until the environment again allows activity; or must in anticipation have left the freezing region; or failing these, must die.

This makes a limited list of choices, but they are met in varied ways. The severity of the temperature challenge increases, of course, with northward distance from the tropics; and correspondingly fewer kinds of organisms have succeeded in finding a satisfactory response. The biota thus becomes progressively poorer in species toward the poles until, in the arctic regions, only a few kinds of organisms have managed to achieve adaptations enabling them to survive the extreme fluctuations of the physical environment.

The effect of seasonal change in climate on population behavior is particularly clear and well studied with insects. Most insects develop rather rapidly, so that the life of an individual from egg to adult requires only a few days or weeks, with many generations in a year. With such species the vast majority of individuals are killed by the unfavorable conditions of winter, and the populations are at a minimum density at the time of the onset of favorable conditions in spring. Numbers may be built up very rapidly, so that in a few weeks the population density has increased a thousandfold. In some cases the increase is fairly steady all through the summer, until the frosts of fall stop insect activity. In other cases, the peak is reached in late spring or early summer, and the population begins to decline in numbers despite apparently favorable climatic conditions. Here we suppose that biological factors—predators, parasites, competitors, available food—have caught up with the expanding population and checked its growth. Each insect species in a given region in the temperate zone thus shows a characteristic annual curve of abundance, and much scientific energy has gone into the study of the characteristics and controlling factors of such variations in abundance.

The hazards of life for the individual in such insect populations change tremendously with the changing seasons. In the first spring generation, it may be that a majority of the eggs laid by an overwintering parent will grow into adults. The hazards (environmental resistance) increase with each succeeding generation until, when the population reaches a plateau of stable abundance, an average of only two of the hundreds of eggs laid by each female will live to reproductive fulfillment. When the population starts to decline, this means that ever fewer individuals are surviving. The conditions of the "struggle for existence," the nature of "natural selection," thus change radically from generation to generation through the year, with consequences that the geneticists are only now beginning to explore. To be sure, many insects in temperate latitudes have only one generation a year, and population fluctuations in such cases are like those of vertebrates, matters of long-term cycles.

GETTING THROUGH THE WINTER

The hibernating mechanisms of insects, whether of the sort with one generation a year, or of the sort with many generations, are varied. Most commonly, perhaps, insects pass the winter in the egg stage, the embryo within the egg being in a state of suspended development. A few insects pass the winter as larvae, usually in special situations, like certain mosquito larvae that overwinter in the mud beneath the ice of the ponds. Only a very few larvae have acquired physiological adaptations that enable them to survive actual freezing. Overwintering in the pupal stage is almost as frequent as in the egg stage: suspension of development seems to be particularly easy for these quiescent states, and pupae, like eggs, often occur in protected situations, such as in the soil. A few insects hibernate as adults which rest, completely inactive, in some protected situation, such as rot holes in trees. With these there is usually some special physiological mechanism whereby the insect is provided with fat reserves which maintain life processes through the months of dormancy.

The common insect method of hibernation in the egg stage is used also by many other invertebrates, and is analogous with spore formation in micro-organisms and "lower plants" and with the common method among annual plants of passing the winter in the seed stage. In organisms in which the life span of the individual covers several years, other methods of surviving the hard times of winter must be found.

Plants for the most part suspend business operations, withdrawing sap from circulation and dropping their useless leaves. With some, all of the exposed parts of the plant are allowed to die, the organism starting life anew in the spring from tubers or bulbs below ground.

The amount of cold that various kinds of plants can withstand is interesting, though not much is known about it. Many tropical trees are killed by frost, and some, like the mangosteen, cannot withstand cold weather, even though temperatures remain well above freezing. Horticulturists in Florida and California are rapidly accumulating an empirical knowledge of what they can and cannot grow, even though they seem not to be spending much time trying to find out the reason why.

It is among the larger animals, mammals and birds, that winter adaptations are particularly varied, and particularly subject to study. Many mammals in the temperate zone remain active all winter, but change their habits or their fur or both. Others go into hibernation, into a winter sleep that varies greatly in its profoundness in different species.

Thus the dormouse goes into "a profound state of torpidity", quoting the *Encyclopaedia Britannica,* not "Alice"; its sleep may last for six months, with breathing so slow as to be almost imperceptible; "it becomes so cold and rigid that it can be rolled like a ball across a table." Ground squirrels and chipmunks hibernate, but with stores of food in their burrows for occasional snacks. Tree squirrels store food, but remain active enough all winter. In bears, the young are born during hibernation, so that mother has to keep up some activities, though relying on her fat reserves for food.

BIRD MIGRATION

With birds the commonest adjustment to seasonal change is migration. This, as Ludlow Griscom has remarked, "is perhaps the most distinctive phase of bird-life, and in its greatest development offers one of the most remarkable phenomena in the animal kingdom." It could be (and has been) the subject of many books, and of much theorizing. The migration routes of a great many species have been carefully charted, and the accumulation of factual observation by bird students is enormous, but it remains an essentially mysterious phenomenon. What was the origin of the habit? What stimuli originate and control the mass movements? What governs the uncanny sense of direction of the birds? There is no general agreement on the answers to these basic questions, which means we don't know.

By bird migration we mean a rather special sort of thing— periodic movement between different regions of the earth's surface in correlation with the seasonal cycle. "Migration" as a common English word covers any mass movement of population, whether cyclic or not, and for the most part such movements are not cyclic, but rather a product of the vicissitudes of climate and population density, as with the waves of human stocks that swept over Europe toward the dawn of history, or the mass irruptions of lemmings charted by Elton, or the devastations of a locust horde eating its way to suicide.

Even among birds there are various types of mass movement that do not show seasonal periodicity—changes or extensions of range related to changing biological or physical conditions. But many kinds of bird activities tend to fall into annual rhythms conditioned by reproductive behavior, for it is a very general characteristic of the class that the sexual organs, ovaries and testes, become functional only once a year. This is true even in the tropics, for purely tropical birds are apt to have a special breeding season, during which they may assemble in particular areas or special sorts of habitats. The whole subject of bird movements and bird behavior in the tropics has been inadequately studied, as ornithologists well realize: a situation that reflects the fact that the majority of ornithologists live in the temperate zone. More precise studies in the tropics may well furnish some of the missing clues to bird behavior in the temperate zone. About 85 per cent of the different kinds of birds are purely tropical, while almost 100 per cent of the ornithologists have habitats that range but little south of Miami, Florida.

The breeding areas of birds showing interzonal migration are in temperate latitudes, and these breeding areas are clearly defined for each different kind of bird. Many "species" sort out nicely into different "geographical races" over the United States during the breeding season, get all mixed up together during their winter southern migrations, and then sort themselves out again as they come north each year. They thus live in reproductive isolation, defined in geographical terms, even though flying together for a good part of the year.

Breeding in the temperate zone must have definite biological advantage for these species, perhaps because they get away from the overcrowded environs of the tropical residents. They find when they get north hordes of insects, seeds and fruits released afresh each spring from winter's inhibition. The birds, in the fall, are forced south by the annual failure of this food supply. The temperature control mechanisms of birds are efficient enough, and most of them can stand the cold as well as any mammal: witness those that have learned to stay in our cities where they are fed regularly, or experience with tropical caged birds, such as parrots, that seem better equipped to cope with the cold than their owners.

No one, however, has yet suggested why some of these birds should undertake the fantastic annual excursions that they are known to make. Thus the upland plover makes a nonstop flight across the tropics to winter in Argentina, Paraguay and Bolivia, as does the Hudsonian godwit. Many of the sea birds wander all over the earth, to return annually to some particular island for nesting. The various species also have characteristic routes, some by land through Mexico and Central America, others straight across the Caribbean, others by island-hopping down the Lesser Antilles.

The migration routes have such a fixed character that they must long have been characteristic for each species, as much so as the colors of its feathers or the notes of its songs. Their development, then, has taken place in the course of geological history, and present patterns are probably as much the results of the climates and geography of the alternating glaciations of the pleistocene as they are of present climate and geography. Which doesn't help much, because while we know a deal about the timing and extensions of the pleistocene glaciations, we can only imagine or guess at the related climatic shifts that must have taken place over the surface of the earth.

Other aspects of migration can be subject to experimental study: the control of the reproductive cycle through hormonal action, the initiating environmental stimuli for such action, such as length of day or temperature, and the correlations between environmental stimuli, reproductive physiology and overt behavior. Much work of this sort has been done, but a great deal more remains to be carried out before we can make many generalizations about behavior mechanisms.

CHAPTER XIII

Biological Geography

IN my first outline, this chapter was headed the "geography of populations", as a sort of logical extension of the discussion of population behavior. I still like the sound of that title, but it would be misleading, because I want to discuss not only the geography of populations, but that of higher groups, of the abstract categories of the systematists, and to try to give a general sketch of the geographical implications of natural history.

Naturalists have always been somewhat preoccupied with the distribution of animals and plants—with the geography of species and of the higher systematic categories—but it is only since the time of Darwin that this knowledge has come to form a special field of science. This is because the distribution of organisms is intimately bound up with the general problems of evolution. The orderly geographical arrangements of organisms, explicable only in terms of their development through geological history, formed an important element in Darwin's argument for the reality of the evolutionary process. His co-discoverer of the theory of natural selection, Alfred Russel Wallace, interested himself particularly in geographical distribution, and his two volume work, *The Geographical Distribution of Animals,* published in 1876, remains the classic of the field.

THE THREE POINTS OF VIEW OF BIOLOGICAL GEOGRAPHY

The study of the geography of organisms can be approached from at least three different points of view. First there is the problem, which interested Wallace more than any other, of delimiting and describing the various regions of the earth's surface in terms that are biologically significant. The political divisions created by the events of human history are meaningless from this point of view, and the relationships of the conventional continents and island constellations of the geographers must be re-examined in the light of the affinities of their inhabitants, of the similarities and differences among their biotas. The same problem exists for marine organisms in relation to the seas and oceans of the geographers.

Then there is the problem of the distribution of different sorts of organisms. The geography of birds or reptiles or fish or fungi must be studied in the light of the geological history of each group, of its methods and powers of dispersal, of its physiology and adaptation to the various sorts of climatic environment.

There is still a third kind of biological geography, the kind that was uppermost in my mind when I started to plan this chapter, the geography of populations. It is useful to think of the distribution of a population in the dimensions of both time and space. A given population, a particular species of organism, presumably starts somewhere, in some definite place, and, as it becomes established, spreads outward with a speed and direction determined both by the characteristics of the kind of organism and by the nature of the surrounding physical and biological environment. It must eventually encounter barriers, sometimes insurmountable, sometimes to be passed after a long interval through the action of chance, sometimes to be conquered by a slow process of adaptation within the species itself, or by the eventual change of the environmental barrier.

The environment, in geological perspective, is never stable; conditions may in time become unfavorable for a given species in the very region where it originated, so that it may die out there, though becoming abundant elsewhere, perhaps in remote areas. The original population, as it expands, may become divided by barriers into a few or many more or less isolated units, which may start out on independent lines of evolution, forming new species. Or the population may remain a connected whole, yet gradually change in character so that species A, with the passage of geological time, becomes sufficiently different to warrant a new name as species B, although continuously occupying the same geographical area. Or, if it fails to meet the challenge of changing conditions, the extension of the population may gradually contract, perhaps leaving little islands of relicts, or perhaps fading into extinction.

We make graphs of everything these days, and I like to visualize the distribution of such hypothetical populations in terms of a sort of three-dimensional graph. At any given time the distribution in space can be plotted on a sheet of paper as a map, an amoeba-like splotch on the cartographer's outline, showing how the species reaches up to Massachusetts and down to Virginia and across into Ohio, with arms extending into the valleys of Kentucky. With the additional dimension of time up or down from the paper, our blotch would be seen to expand and contract, push out arms tentatively here, boldly and successfully there, reaching backward finally to the pinpoint of its origin, or forward to the pinpoint of its extinction. And, amoeba-like, we might find it dividing now and then along this time dimension—but that is a problem not so much of geography as of the origin of species.

But we should start this discussion with the first geographical problem, that of delimiting the various regions of the earth's surface in biological terms.

THE MAJOR BIOTIC REGIONS

Wallace based his zoogeographical studies largely on birds and mammals, because those were the only groups sufficiently well known in his time to serve as a basis of generalization. The striking characteristic of birds and mammals is the uneven distribution of higher categories, of genera, families and orders. For instance, the marsupials today are limited to Australia, where they have proliferated in extraordinary variety, and to America, where we have the common opposum with a few rather inconspicuous South American relatives. South America is inhabited by a whole series of families of mammals and birds—sloths, armadillos, cebid monkeys, hummingbirds, and so forth—found nowhere else in the world except for extensions into North America. The Australian fauna is equally queer. The inhabitants of Africa south of the Sahara are less conspicuously different, but still represent a different fauna from those of India and the oriental islands. In temperate latitudes, the faunas of Europe, Asia and North America are very similar, differing utterly from those of the southern temperate zone, the tips of South America, Africa and Australia.

The genera and families of seed plants have had, on the average, a rather longer geological history than the corresponding categories of birds and mammals, and have thus had time to reach a more uniform geographical distribution in regions of appropriate climate. They also have, in seeds, a very efficient means of dispersal. The regional characteristics of plant groups are thus not as sharply or conspicuously defined as those of the vertebrates, and students of "phytogeography" put less emphasis on faunal regions than do students of "zoogeography." The plant people (I think this is a safe generalization) are apt to put more emphasis on methods and rate of spread of specific populations, on climatic adaptations, and things of that sort.

But Wallace's regions, which have been very generally accepted by zoologists, serve also as a framework for discussions of plant geography. Following a scheme first proposed by Sclater, Wallace recognized six general regions: the Palearctic, including Europe, temperate Asia and north Africa to the Atlas Mountains; the

Ethiopian, including Africa south of the Atlas; the Oriental, extending from the Himalayas to South China and Borneo and Java; the Australian, including New Guinea, Celebes and other islands; the Nearctic, including North America as far south as the highlands of Mexico; and the Neotropical, including South America, the Antilles, and Central America north to central Mexico. Wallace and more recent students have divided these major regions into many subregions; but in fixing boundaries for the subregions there has been little agreement.

The animals and plants of Europe and North America are essentially very similar, and a great deal of ink has been used in discussion of whether the Nearctic and Palearctic regions should be separated, or united as a single circumpolar Holarctic region. The genera are the same in most cases, though the species on the two sides of the Atlantic are almost always distinct. It is quite certain that Alaska and Siberia have been repeatedly connected by a "land bridge" in the area of the Aleutian Islands in the geological past, and during interglacial periods, the climate allowed for an easy exchange of biota between the two hemispheres.

The differences between the Nearctic and Palearctic regions are due largely to intruders from the south. The North American fauna in particular includes a great many types of animals belonging to peculiar South American groups, not found in other parts of the world. Our field mice belong to a South American family (the Cricetidae), quite different from the Old World mice, which are represented on our continent only by the house mouse and the domestic rats that have come over in historic times. Opossums and raccoons are other representatives of South American families not found elsewhere in the world (except that the Asiatic pandas are distant cousins of our raccoon). The migratory birds include a host of South American types that have gradually spread their ranges north with the retreat of the glaciers. Europe and temperate Asia, similarly, include among their inhabitants many Oriental and a few

African types that do not correspond to any element in the North American fauna.

Thus it is very convenient to separate the temperate zone land areas of the two hemispheres as distinct geographical regions, if only because they are subject to different influences from the teeming tropical regions to their south. The general pattern of animal distribution, however, is more easily understood if we keep in mind the essential continuity of the northern temperate land mass.

It is instructive, from this point of view, to look at a world map on a polar projection. The esential continuity of the land mass surrounding the north pole is then clear, and this mass is seen to have three divergent extensions into the equatorial and southern regions: first, South America, second, Africa, and third the Asiatic tropics extending through the East Indies to Australia. Australia, completely separated by sea, and South America, attached only by the slender thread of Panama (frequently broken in times past) are then seen in their proper perspective of remoteness.

They both are very peculiar areas from the European or North American point of view. The Holarctic naturalist, going to either, is making a voyage back in geological time, because in either one he will find animal types not known from his own area since, say, the Miocene. Whether South America or Australia is the more peculiar is a matter that might be debated at length. Australia would seem to win at first sight, if only because of the extraordinary evolution of the marsupials there. But in any statistical list of the peculiarities of the two continents, South America wins, because its great forests are inhabited by an almost endless series of animal types found now in no other part of the world.

The great paleontologist, William Matthew, wrote a very persuasive book *Climate and Evolution* in which he argued that the major evolutionary advances in animal development must have taken place on the northern land mass, chiefly perhaps in Central Asia, because the distribution of mammals, both in the geological past and at the present time, could be explained in terms of successive waves of migration out from this center. These new types replaced their less efficient brethren first in this Holarctic area, extending into the tropical regions more gradually, as time and geological land connections allowed. The most ancient and the most primitive types thus survived the longest in the most remote areas, Australia and South America.

The isolation of Africa by the Sahara, of the continental Orient by the Himalayas, and of the insular Orient by water barriers, is less effective and geologically more recent, than the isolation of either South America or Australia. Africa and the Orient have, in consequence, fewer peculiar primitive types surviving in their faunas.

Matthew's thesis is that the new and more efficient mammal types, such as horses, dogs, cats, primates, have had their origin in the great Asiatic land mass. Evolution has not been arrested in either Australia or South America; it has followed a different, and less radical course. The evolution of the marsupials in Australia has been, in fact, extraordinary. They have taken over all of the mammalian roles that in other regions are played by very diverse structural types: there are rabbit-like marsupials, wolf-like marsupials, arboreal squirrel-like marsupials, and the kangaroos which take the place of the large grazing mammals of other regions. But all of these animals have remained marsupials, have kept a basic structure that is less efficient, that cannot hold its own in competition with the more modern mammal types that have meanwhile been occupying much of the rest of the world.

In South America, local evolution among mammals has chiefly taken the form of adaptation to the great Amazonian forest. Prehensile tails, for instance. Only one mammal in all of the world outside of the Neotropical area has a prehensile tail—one of the Australian marsupials. But in South America all kinds of mammals have developed such tails, an independent invention with each. There is an arboreal porcupine with a prehensile tail; a carnivore, the kinkajou, a relative of the raccoon; an anteater, belonging to the peculiar order of edentates; and several marsupials. All of this besides the monkeys, whose marvellous grasping tails are well known. Any time you see a monkey hanging by his tail, you can be sure he came from South or Central America, not from Africa or the Orient.

The thesis of Matthew, that the major developments of vertebrate evolution occurred in temperate Asia, has been challenged by many students. Most plausibly, Philip Darlington has shown that the evolution of the primary groups of cold-blooded vertebrates, fish amphibia and reptiles, seems to have occurred in the tropics of the Old World, rather than in the temperate latitudes. In a sense, though, Darlington's work, which puts the center for cold-blooded vertebrates a little south of the center for mammals, re-enforces Matthew's principle thesis, that evolution has been most rapid in the largest land mass, and that the spread outward has been in accord with land geography not strikingly different from that shown by the continents today. Perhaps the cold-blooded vertebrates, for physiological reasons, proliferated most greatly in tropical latitudes, and mammals in temperate latitudes.

The conspicuous birds and mammals that are characteristic of the various regions are pretty well known. There are parallel differences among the cold-blooded vertebrates and the insects. The seed plants, as I remarked, show less striking regional characteristics, though there are some outstanding peculiarities, such as the limitation of the bromeliads (the pineapple family) and the cacti to the Neotropical region and the great development of the gums, Eucalyptus, in Australia.

With many groups of microscopic organisms, both of animals and plants, geography loses all meaning. With bacteria, protozoa, and algae, for instance, the same species may be found in all parts of the world where environmental conditions are suitable for that particular species. This, of course, is because the resistant spore stages may be carried indefinitely as dust in the atmosphere. Strictly parasitic species are chained by the geography of their hosts, and specialized habits or poor dispersal mechanisms limit the distribution of many free living types. But phrases like "Australian bacteria" or "Neotropical protozoa" have no meaning comparable to that of the same adjectives coupled with mammals, birds, butterflies or molluscs.

THE GEOGRAPHY OF AQUATIC ORGANISMS

Geography, too, has a different meaning with marine organisms. The pelagic types, the organisms that drift or swim in the open sea, do not encounter the insuperable barriers of land organisms, and many of them, from whales to microscopic crustacea, are liable to be found in any part of the world. There is a remarkable similarity in the inhabitants of the seas of the two polar regions. Various theories have been proposed to account for this. Some organisms may be able to pass from one polar sea to the other through the cold waters of the ocean depths.

Shore (littoral) organisms are of course subject to the same sort of limitations to distribution as land organisms, except that land is the barrier instead of water, or open seas instead of mountain ranges. The inhabitants of shallow tropical waters, especially, may be confined to a quite limited range by their environment and their habits.

Fresh-water habitats present special distribution problems because of their discontinuity and their transience. No fresh water is permanent in the sense that the seas and the continents are permanent, and only the great river systems and a few of the lakes, like Baikal and Tanganyika, have persisted through significant stretches of geological time. The Great Lakes of North America look impressive and permanent enough, but they were left by the retreat of the last Pleistocene glaciation, a mere moment of geological time.

Fresh-water organisms, then, immediately raise the question of mechanisms of dispersal: how did they get where they are? A rain pool, in a few days, starts to teem with life. If we examine the catalogue of its inhabitants, we find that they all have either some resistant spore-like stage produced in immense numbers and broadcast with the wind so that some are liable to fall into suitable water when and where it accumulates; or that they have some life history stage that is able to survive in earth or mud from one inundation to the next; or that they have life histories in which terrestrial stages alternate with aquatic stages, the terrestrial form taking over the function of dispersal.

Life presumably started in the ocean, in salt water, later acquired adaptations permitting survival in fresh water, and still later adaptations for land. The fresh-water organisms of today, however, do not directly reflect this sequence. Many of them—the fish, for instance, most of the molluscs, and things like fresh water sponges—are clearly derived from marine ancestors; but a host of others clearly had terrestrial ancestors. These represent what the biologists would call a secondary invasion of the aquatic habitat. A few things with terrestrial ancestors, like the marine mammals (whales, dolphins, sea otters and so forth), have re-invaded the sea.

Insects, we think, arose as a terrestrial group, and they have certainly proliferated tremendously on land. A very wide variety of these land insects have independently re-invaded fresh water, for some stage or other of their life history. The adult stage, however, if it is not purely terrestrial, has at least retained the ability to get out of the water from time to time and fly off to lay its eggs in whatever suitable water accumulations it can find. Some of these insects—some dragonfly species, for instance—are very powerful fliers, and are thus able to colonize the most remote and disconnected bits of fresh water. Some of the mosquitoes also have great powers of dispersal. A few other aquatic organisms, like the water mites, get free rides to new habitats by attaching themselves to these insects. Other kinds of organisms get wide dispersal through mud sticking to the feet of wading birds. The aquatic amphibia and reptiles, of course, get about through terrestrial stages or ability to survive under either water or land conditions. The amphibia, frogs and salamanders, are presumed to represent an evolutionary stage that has never learned to get completely free of the aquatic habitat: their fresh-water adaptations are primary, unlike those of the insects.

The fresh-water fish of the world fall into three groups, as Darlington has emphasized: those families which possess an ancient physiological inability to survive in sea water, which binds them to the land as securely as any known animals; other families that live chiefly in fresh water, but that sometimes enter the sea and that can survive there for a limited time; and still others that are semi-marine or migratory, or derived from such semi-marine families. The first group are the most interesting to zoogeographers, since the existence of related forms in different water systems presumes some previous connection between those systems. There are, for instance, no fish of this group on the islands of the West Indies; and none in the Australian region, except for a very ancient type of lung-fish.

Observations like these form the raw material of zoogeography in the conventional sense, the geography of biotic regions. The discussion of fish brings us to the second aspect of biological geography: the study of the distribution of different sorts of organisms. This involves consideration of the different kinds of barriers that exist, and of the different dispersal mechanisms that various organisms have adopted.

THE DISPERSAL MECHANISMS OF ORGANISMS

Among animals, dispersal generally depends on the ordinary form of locomotion of the organism—flying, swimming, walking, crawling—and the barriers are the sort of thing that cannot be passed by this means of locomotion. The effectiveness of barriers, however, depends to a great degree on the behavior patterns of the species. Thus most birds can fly, and the migratory species bear convincing witness that apparently very fragile birds can have tremendous powers of endurance in flight. Yet many birds are chained by their habits. The two banks of the Amazon river in a whole series of cases are inhabited by different subspecies of the same kind of bird, indicating that the river is a real barrier to the birds. They undoubtedly could fly across it, but they don't. Similarly some mammals, birds and insects apparently pass freely across particular mountain barriers, while others, that look just as capable of making the passage, remain settled each in its own little valley.

The mechanisms of dispersal are particularly interesting in sedentary organisms. Thus seed plants have developed a whole series of devices for getting their seeds carried from one place to another. We are all familiar with a variety of such adaptations: the stick-tights and burrs that we pick off flannel trousers; the floating devices of milkweed and maple seeds; the fruits that advertise to birds, get eaten, and have the seeds deposited at the next stop.

The dispersal mechanisms of parasites form another special field of study. Parasitic life is a fine thing insofar as the host does all the hard work, but it presents the special problem of how to get from one host to another: a problem that increases in complexity with increased specialization of the parasitic habit. The complicated parasite life histories outlined in an earlier chapter serve this function of dispersal—a function that is clearest in the vector relationship, as with malaria and Anopheles, yellow fever and Aedes, typhus and lice.

The zoogeographers also have to consider another sort of dispersal problem, that of how the organisms existing in a particular place got there. This becomes a particularly involved game in the case of isolated situations, such as oceanic islands or landlocked lakes.

The island problem is a particularly good puzzle. Since complete proof of what happened in each case is hopelessly buried in the geological past, the puzzle can hardly be unequivocally solved, and thus serves indefinitely for the exercise of the imagination. In the case of truly oceanic islands, such as the Pacific atolls, all of the inhabitants must have got there, somehow, over the sea. In the case of clearly continental islands, such as the British Islands, the sea barrier is a very recent thing geologically, so that the ancestors of the present inhabitants may have got there by walking over land. But a great many islands fall somewhere between these two extremes, and the question of their continental connections is open to endless debate.

Certainly a great many organisms are found in situations of the sort that make it extremely difficult to imagine how they ever got there, and naturalists have dedicated a great deal of ingenuity to thinking up possible mechanisms. A favorite controversy is over life rafts. Anyone with experience with a great tropical river has watched huge sections of bank cave in and go drifting downstream toward the sea. Almost anything might get passage on such a raft and be conveyed, say, from the upper Orinoco to some Atlantic island. The hazards are immense if the naturalist is trying to get some delicate organism, say a frog, carried any distance. The life rafts don't hold together very long; the organisms would be killed by salt water in the open sea, or after stranding on a beach; either a pair or a fertilized female would have to be taken, and so on. But if the proposition is that such an event might have happened once in a million years, it is removed from the area of experimental testing and becomes a matter of my ingenuity against yours in disputation.

If some species or other of frog is found living on an island, it must have got there somehow. If it didn't arrive on a life raft, or wasn't blown clinging to a branch in a hurricane, the island must at some time have been connected by land with some region where such frogs already existed. We know that the geography of the earth has been constantly rearranged through geological history, that seas once spread where there are now lofty mountains, so that connecting islands and continents by land bridges is not necessarily an improbable exercise of imagination. But it sometimes makes difficulties greater than those it solves. If we build a geological land bridge to get a particular frog to an island, we may be faced with the problem of explaining why a lot of other things didn't come along too.

THE GEOGRAPHY OF POPULATIONS

To finish this sketch of the problems of biogeography, as outlined at the beginning of the chapter, we have to consider the third point of view, that of the geography of populations. This has increasingly come to interest a variety of different kinds of naturalists. The student of classification, the taxonomist, has found it necessary to give detailed consideration to the geography of his species, particularly in groups like vertebrates and insects where the majority of populations are clearly geographically defined, forming distinct geographical subspecies in different parts of the range of the group. Many of these students have come to the conclusion that it is only through the geographical isolation of one part of a species population that the population can become split into two species, that therein lies the mechanism of the origin of species.

The geneticists have also become interested in the geography of populations, and are producing an increasing volume of studies on the geography of variability, on the differences and similarities of the hereditary material, the genes, in different parts of the range of a given population, or species. This is slow and painstaking work, but it gives promise of making an important contribution to our understanding of the process of evolution.

The ecologists are also much concerned with the geography of populations. They would like to know, among other things, the limiting factors in spread: which, for instance, of the environmental factors limit the northward or southward extension of a particular species. They have forced even the purest of the geographers to think also in ecological terms, to consider not only the cartographic shore lines and contours, but also the total environmental situation within the areas that they have mapped.

Man must have the widest distribution of any land organism, except for some of the ubiquitous bacteria. Some of the animals he has domesticated, and some that have adapted themselves to him, like mice, fleas and lice, would come next, with the dog coming closest to going everywhere that man does.

If we leave this human complex out, the only terrestrial animals that have achieved world-wide distribution are some of the sea birds—which is probably stretching the definition of terrestrial. Their general wanderings over the seas are comparable to the wanderings of some of the marine animals, and would have the same basic explanation—the relative continuity of the seas as contrasted with the dicontinuity of the land masses of the earth.

THE RANGE OF DIFFERENT SPECIES

In general, species that range over a continent are few—leaving out of account organisms like bacteria and the special human entourage. The American puma has an extraordinary range for a mammal, from Patagonia to southern Canada. Taxonomists have split off some 19 geographical subspecies over this range, but these are based on pretty trivial characters. The jaguar has almost as wide a distribution as the puma in America, and has the further distinction of being but very slightly different from the leopard of the Old World, which itself has an extraordinarily wide range over Africa and southern Asia.

I can think of no wild seed plants with ranges comparable to these, except some of the weeds that have trailed along after man. Even the cultivated plants and weeds associated with man are apt to have much more sharply limited distributions than the corresponding animals, because of their more exacting temperature and light requirements. Cultivated species with wide distributions, like maize and wheat, have been adapted to the differing environments of different latitudes by the development of special strains with different growing habits and different requirements for temperature and length of day. This is, of course, true to some extent of domestic animals; but the dependence of a given strain on light and temperature conditions is nothing like as striking as in plants.

At the other extreme, species with very restricted ranges are to be found in almost all kinds of organisms, particularly on islands or isolated mountain peaks, or in isolated valleys or streams. Organisms with sedentary habits, or with habits that chain them to special situations like heavy forest shade, are particularly apt to break up into endless series of isolated populations on islands or in mountains.

A botanist, J. C. Willis, has proposed a theory in a book called *Age and Area* to the effect that the range of a given species is directly related to its geological age. This may have a certain limited application, in so far as with increasing time, a population has increasing opportunities to extend its area of colonization. But sooner or later it comes to insurmountable barriers. And also the process of extinction, of contraction of range, must be just as common in nature as the process of expansion. As a matter of fact, some of the most restricted ranges are characteristic of the most ancient organisms, "relic species" that have lingered on from past geological ages in some inaccessible spot.

Extent of range, and consequent size of population, has come to be a matter of great interest and lively debate with students of evolution. Something new, a mutation changing the character of a population, might stand a far better chance of getting established in a small population than in a large one, where the novelty might be swamped by the overwhelming numbers. But on the other hand, there would seem to be a far better chance that something new, something advantageous, might appear somewhere among the very large numbers, than among the small numbers of a population with restricted range.

This, however, is directly a problem of the mechanism of evolution; though it is difficult at any point to disentangle the subjects of biogeography and organic evolution. Before reviewing current theories about the process of evolution, however, it may be well to interpolate a chapter on adaptations, on the interdependence of organs and functions within the organism, and between the organism as a whole and the environment.

CHAPTER XIV

Adaptations

TO adapt, Webster says, is "to make suitable; to fit; to adjust". Adaptation in its biological sense, according to this same dictionary, is "modification of an animal or plant (or of its parts or organs) fitting it more perfectly for existence under the conditions of its environment". It is thus a pretty broad word, both useful and much used in biology.

All the way through this book I have been concerned with the relations of organisms to each other and to the environment, which involves adaptations of form, function and habit to the extent that one can wonder, what in the organic world isn't adaptive? Julian Huxley answers this by pointing out that there are certain basic functions, such as assimilation, reproduction and reactivity, that are common to all protoplasm, inherent in the nature of living matter, and that these things can hardly be called adaptations. The manifestation of such functions in any given organism, however, may be greatly modified by adaptations, by becoming fitted to the peculiar needs and situation of that organism.

Huxley illustrates this with the example of regeneration of lost parts. The German evolutionist, August Weismann, suggested that the property of regeneration might be an adaptation acquired by organisms that were especially exposed to the loss of limbs or similar damage. Experimentation with all sorts of organisms, however, has gradually led to the development of the concept, which I outlined in Chapter V, that potentially any cell of a complex organism can reproduce the whole organism, and that this potentiality has gradually been lost, inhibited, or modified in different organisms, until in mammals the only traces are the phenomena that we observe in wound healing. Regeneration of a whole earthworm from a cut half, then, is not so much an adaptation to the hazards of earthworm existence, as a manifestation of the degree to which cellular totipotency has persisted in earthworms.

On the other hand, as Huxley points out, there is a curious special modification of regeneration that can be considered as an adaptation. This is the development in many organisms of the faculty of autotomy, or self-mutilation. A lobster, for instance, if caught by a leg, may break off the leg and escape. Many a boy, catching lizards, has found that he is holding only the wriggling lizard tail, which has been cast off by the frightened animal. The mechanism whereby these parts can be cast off, with a minimum of injury to the animal, is a rather specialized adaptation for escaping predators. The subsequent regeneration of the lost organ, however, could not be considered adaptive.

Taxonomists often get involved in arguments as to whether particular characteristics of a given species are adaptive or not. The wing of a butterfly, for instance, is clearly an adaptation for flight. The pattern on the wing may be such as to resemble a dead leaf very closely when the insect is at rest—surely an adaptation for concealment. The patterns and colors on butterfly wings, however, are endlessly varied, and no one has been able to think up adaptive explanations of all of the patterns. Two species of snakes may most readily be distinguished by scale counts, two species of mosquitoes by the numbers of branches on a particular bristle. Again, it is hard to visualize the adaptive significance of these differences.

Some naturalists feel that everything about an organism must be adaptative, modified to fit some peculiar circumstance of that organism, and exercise great ingenuity in thinking up explanations for apparently trivial or useless characters. Other naturalists, in reaction, swing to the opposite extreme and tend to consider all nature a sort of concatenation of accidents.

These last are apt to be armchair, or laboratory, naturalists. The field man is generally so impressed by the fitness of everything, that he sees adaptations everywhere, reaching the extreme in the famous case of the New England naturalist and painter, Thayer, who included a painting of pink flamingos against the sunset in his book on concealing coloration.

In discussing adaptations there is an almost irresistible temptation to describe the bizarre, the striking cases, and overlook the millions of work-a-day adjustments that make it possible for the organic world to function. Since my experience has been largely that of a field naturalist, I shall probably fall into the same error here, of putting emphasis on striking examples. But these striking "wonders of nature" do attract attention, do thus arouse a need for explanation, and come finally by that process to play what may seem a disproportionate role in the development of natural history. Disproportionate or not, their importance in the discussion and thinking of naturalists cannot be denied.

Julian Huxley has conveniently grouped adaptations into three classes: those related to the inorganic environment, those related to the organic environment, and those related to internal adjustments in the organism itself. These, as Huxley points out, overlap and intergrade, but they serve as a convenient frame for discussion.

ADAPTATIONS TO THE INORGANIC ENVIRONMENT

An attempt at analysis of any one of these groups makes one realize at once the bewildering complexity of adaptation phenomena—and their universality. As an example of adaptations to the inorganic (physical) environment, consider aquatic organisms. The density of the medium means that streamlining is necessary if any speed in swimming is to be attained. The swift aquatic vertebrates— fish, reptiles (mostly extinct groups like the ichthyosaurs) and mammals—have all, through adaptation, attained remarkably similar streamlined bodies. Birds of several diverse groups that have taken to swimming on the water surface have developed webbed feet. Animals with terrestrial ancestors that have taken to the water, like the aquatic mammals and insects, have developed adaptations of structure and habit to enable them to get air, or to utilize oxygen dissolved in the water. Marine organisms require special adaptations for life in fresh water, so that the salt content of their body fluids can be maintained at a different concentration from that of the surrounding medium.

The list of adaptations involved in the slow process whereby organisms got out of the water is even more complex: adaptations for support in the lighter medium of air, for the acquisition of water and its retention in the body, for rapid locomotion in this lighter medium by running or flying, for the transfer of gametes in sexual reproduction. I can't do more than attempt to suggest the general nature of the problems that had to be met.

Then each specific kind of habitat involves adaptations for its special conditions, adaptations that are often curiously parallel in different organisms. Thus luminescence is common to many deep-sea organisms. Cacti in the New World and Euphorbias in the Old have developed parallel modifications for existence in the desert, examples of a whole catalogue of adaptations for desert, or "xerophytic" conditions. Forest adaptations include the prehensile tails mentioned in the last chapter along with other adaptations for climbing. The tree form itself might be treated as an adaptation for forest conditions, and called "giantism", which has been adopted by a whole range of plant families from grasses (bamboo) to compositae (plants of the daisy family). The climbing adaptations of animals are of course related to an organic factor in the environment, the trees: but the trees, for the animals, are filling the physical function of support; tree growth, though itself organic, provides the physical skeleton around which the structure of the forest community is built. The climbing or liana habit, of plants, and the epiphytic habit, are then further examples of adaptation to the physical conditions of the forest habitat.

ADAPTATIONS TO THE ORGANIC ENVIRONMENT

Adaptations related to the organic environment would include all of the special modifications developed by organisms for the purpose of living together, whether for mutual advantage as in the case of symbiosis, or for exploitation as in parasitism and predatism. It is among this class of adaptations that the bizarre is most apt to attract attention: examples like the Australian orchid Cryptostilis which, as Huxley says, practises an ingenious variety of prostitution, for the flowers resemble the females of a particular fly both in form and odor, and are pollinated when the males of this fly try to copulate with the flowers.

The chapters on symbiosis and parasitism include various remarks on the sort of adaptations involved in those relations, so that there is no point in repeating here. I would, though, like to emphasize again the difference in direction of adaptation in the two relationships. Where two organisms are symbiotic, each is directly adapted to life with the other. On the other hand, a parasite or predator is adapted to living in its host, or to catching its prey; but the direct adaptations of the host or prey are toward protection. The antibody system of mammals is an adaptation to protect the host against parasites; the fleetness of grazing mammals is an adaptation to escape from predators.

Of course, if you keep on digging, you find that adaptations like those between predator and prey may have a reciprocal aspect. A high reproduction rate in prey and a low one in predator can certainly be looked on as a mutual adaptation. But this is an aspect of the general interdependence of natural phenomena. By going a little further, it can be said that physical conditions at the surface of the earth are adapted to life, as well as that life is adapted to these physical conditions: a line of thought that has been explored by L. J. Henderson in his book on *The Fitness of the Environment.*

This, however, is digressing from the consideration of adaptations to the organic environment. Three general classes of such adaptations have come to form the subject of an extensive literature, and have played an important part in the development of theories of the mechanism of evolution, so that they merit more than passing attention here. These are the topics of concealing coloration, warning coloration, and mimicry.

CONCEALING COLORATION

I think that no naturalist who has worked in the tropics can doubt the reality of concealing coloration. It is common enough in the temperate zone for animals to blend so thoroughly with their environment that they are difficult to see, but temperate zone biologists, particularly if they spend most of their time in the laboratory, are able to shrug this off as an interesting but unimportant phenomenon, an accident of convergent evolution, a result of the operation of some irrelevant physiological process. Green caterpillars, they will admit, are difficult to see on the grass; but the caterpillars are constantly ingesting large quantities of green chlorophyl, and why shouldn't this accumulate in the epidermis? They can also collect a fairly impressive list of animals or plants that look like other animals or plants even though they live in quite different places; and an even more extensive list of animals that seem to get along very well in the world without looking like anything in particular. House flies, for instance, are biologically "successful", even though they bear no resemblance to the walls that they sit on.

But such a biologist is badly shaken when one leaf on a branch

he is holding starts to walk away; when the flying butterfly that he
has been trying to catch simply disappears, though he could swear
he saw the exact spot where it lit; when a patch of lichen on a tree
trunk gives him a severe burn and begins to crawl. It is possible to
shrug off all of the pretty pictures that naturalists have used to illustrate their travel books as mere extremes of Thayer's picture of
flamingos against the sunset. But I cannot imagine anyone working
in a tropical rain forest or on a coral reef and still doubting the
purposive nature of the blending of organisms with their environment. Perhaps I had better dodge by pointing out that by "purposive" I mean directed toward a particular end, not random, incidental, or arrived at by chance.

NATURAL SELECTION

Concealing coloration leads inevitably to the discussion of
natural selection: the two topics have been completely interwoven
since that day in 1859 when the *Origin of Species* was published.
The possible process is hard to visualize by starting with the extreme
cases, like those in which the folded wings of a butterfly copy in
every particular the venation, the mould spots, the tattered edges of
a dead leaf. But the blending of animals with the environment is a
very general phenomenon, and every stage in specialization can be
found. There is a general tendency of animals living among green
plants to be green, for those living among plant litter to be brown.
Fish are very generally dark colored above, light colored below, so
that they blend with the surface against which they are seen in either
case.

These generalized resemblances can be followed by every sort
of intergrade to the perfect and detailed copying described in the
"oh my!" books on nature. It is easy to visualize, as Darwin did, the
steps that might have brought this about: the predators becoming
constantly more keen eyed as their prey became more inconspicuous, ill-fitting individuals being caught first, those better resembling

their background surviving to propagate the species. This would lead, as surely and inevitably, to the final perfect product and to its maintenance in this perfect state, as does, with man, the constant selection of wire-haired terriers for their particular points.

This is all very nice, but is it true? Does selection really happen that way? The discovery of mutations, of the fixity of the hereditary material in the germ plasm except for apparently haphazard, accidental and occasional shifts of the genetic material, seemed at first to give a death blow to the theory of natural selection. With a mixed lot of peas the experimenter could select for largeness and roundness, or any other character, until he had reached a "pure line", a strain with homogeneous genetic material, but after that all further efforts were unavailing until or except for the accident of some mutation or other. The early geneticists very confidently put natural selection away in the attic, and announced their explanation of evolution in quite other terms.

The fallacy seems perfectly obvious now. The genetic studies of Mendel, De Vries, and Morgan had nothing to do with the trueness or falsity of natural selection. They were explaining a mechanism of heredity, a means whereby similar characters could be transmitted from parent to offspring, variation within the population maintained, and changes in new directions occasionally added. Natural selection would operate at a different level, governing the distribution of this genetic material in the populations, determining the survival or elimination of new mutations, and their rate of spread. Realization of this difference in level has brought the geneticists out to the field, where they are involved in the laborious task of determining the frequency of mutations, and the actual distributions of genetic materials among wild populations.

Students of natural selection, in the course of these events, have been forced to stop saying "oh my!" at the butterfly looking like a leaf, and to learn the operation of calculating machines and slide rules. Two mathematicians have led the field, Sewall Wright in the United States, and R. A. Fisher in England. They have shown that even if an individual organism with a slightly different genetic make-up has only a very slight advantage over his fellows—a chance of one in 99 of surviving instead of one in 100—this slight selective advantage may have a profound effect on the spread of those slightly favorable genetic materials through the population as a whole. It is all explained in Fisher's book on *The Genetical Theory of Natural Selection,* and in Dobzhansky's on *Genetics and the Origin of Species*—though neither book is likely to be classed as light reading.

Natural selection faces other difficulties. We look at the world through our own sense organs, and necessarily interpret the world in terms of our background of experience, which gives us a constant bias. The butterfly may look exactly like the leaf to our eyes; but if we photograph it with ultra-violet light, it may look much less like the leaf; and the animals that prey on the butterfly may, in seeing, use a different part of the spectrum from the one that we use, including light of shorter wave length. This is not theoretical: many cases of protective resemblance do lose force if examined with ultra-violet light, and there is much evidence that many possible predators discriminate farther into the ultra-violet than we do.

Then, too, what good does it do the caterpillar to look exactly like the stem of the plant on which it feeds, if the chief enemies of the caterpillar are parasites that locate their hosts by smell? Maybe, of course, nature is full of protective smell resemblances, but we are handicapped in judging by the unique character of our own particular spectra of sense perception.

The dangers of human interpretation of protective adaptations is nicely illustrated by MacGinitie's observations on the action of the "ink" cloud produced by a frightened octopus. This ink discharge is completely analogous to the human device of a smoke screen, and one would automatically assume that it served that function for the octopus. From observing octopuses and moray eels (which specialize in octopus eating) MacGinitie concluded that the actual effect of the ink discharge is to paralyze the sense of smell of the moray (or other enemy). Sometimes the moray would not recover its sense of smell for an hour or two after the ink had dissipated—the time interval depending on the dose of ink received. To quote MacGinitie: "the sense of smell of the eel has been so paralyzed that we have often seen the moray actually put its nose against the octopus and not know the octopus was there."

The theory of protective resemblance has been attacked most strongly by the American naturalist, W. L. McAtee. Among other things, he has examined the stomach contents of thousands of birds and analyzed the results in detail. He has found that supposedly protected insects are destroyed by birds in numbers that seem to be proportional to their abundance as compared with species without special protective coloration. The answer of the selectionists, of course, is that the methods of sampling used in work like that by McAtee are not exact enough to reveal the very slight differences that would give a significant advantage to the protectively colored insects.

Finally, a common objection to the theory of protective coloration brought about through natural selection is that the supposed advantages of the protection must be greatly exaggerated, since a host of organisms manage to get along very successfully without protective coloration. This, as Huxley has pointed out, is like saying that electric refrigerators must be useless because many people who could perfectly well afford them go without—an analogy that has more force for an Englishman than for an American.

This objection is really a special case of the general proposition that organic nature must represent a perfected system easily comprehensible in human terms, instead of a hodgepodge of all sorts of adjustments that somehow works after a fashion. It is easy to be impressed with the wonders of nature, with the beautiful balance and adaptation. But it is also easy, from the human point of view, to make out a case that nature is a very badly arranged affair indeed.

Often in watching the events of a forest, I think that if given a chance, I could arrange the whole business much better: that the place is full of empty niches, badly wasted food supplies, organisms ill-adjusted to their environment. At one moment, the keen struggle for existence, the constant pressure of natural selection, the whole beautiful balance of the community seems very clear. At another moment and following a different line of observation, it seems that the place has so much food, so much light, such a readily available supply of water, that almost anything could survive and that almost everything does, after the manner of an opulent but sloppy household where one would think nothing could be accomplished, but where the people seem to thrive and after their fashion get things done.

So the mere fact that all of the populations in nature do not have the electric refrigerators of protective coloration worries me not at all. The interrelationships, the advantages and disadvantages, the adaptational possessions, of each population are different. We may never understand them all, but it is our own nature to keep on trying.

WARNING COLORATION

I got directly into a discussion of natural selection from protective coloration without considering the two related topics of warning coloration and mimicry, though the discussion pro and con about natural selection is usually built around all three subjects.

Warning coloration, like concealing coloration and all of the rest of these adaptational phenomena, is most apparent under tropical conditions, where a maximum diversity of kinds of life is packed into a minimum of space. There are, under such conditions, a great many very conspicuous organisms that seem intent on advertising their presence. They are, with surprising frequency, colored red; and on investigation they seem very generally to have some special form of protection, like a sting or a nauseous secretion. Now the theory is that if an animal has some such special form of protection, it is to the advantage of the animal to advertise itself as boldly as possible, so that possible predators will recognize it and leave it alone. The predator, for instance, will learn, after a minimum of tries, that a conspicuous wasp is not to be fooled with; whereas, if the wasp looked like some common harmless and tasty sort of insect, it would take the predator a much longer time to learn to distinguish wasps and to leave them alone; more wasps would thus be killed in the course of the learning process, and the great defensive advantage of the sting would be lessened.

The theory of warning coloration is not all plain sailing. Coral snakes are very poisonous, and are conspicuously marked with colored transverse bands. But they are nocturnal animals which spend the day hidden in crevices in rocks and like places; and at night the conspicuous coloring would seem to lose its value. And other poisonous snakes may show protective coloring instead of warning coloring; the fer-de-lance, for instance, which blends perfectly with the litter of the forest floor.

Related to warning coloration is the bluff in behavior or in colorings shown by many animals. A harmless snake may show belligerent behavior (often scaring me, at least). An inconspicuous and innocuous moth, disturbed, may open its forewings and disclose hindwings marked with a brilliant "eyespot." The moth certainly acts as though it hoped the eyespot would frighten the disturber away.

MIMICRY

Warning coloration and bluff lead naturally to the fascinating and extensive subject of mimicry. Mimicry, among zoologists, has come to have a special meaning, applying to the cases in which one animal looks very much like another animal, not closely related. It was first described by Henry Walter Bates, who noticed in the course of his Amazonian explorations, that many butterflies of different families (as shown by anatomical structure) looked very much alike. Where several such unrelated species showed the same color pattern on their wings, he noticed that all would have similar habits, leading them to fly in similar situations, that one species would be much more abundant than the others, and that this common species, the model, would have some special characteristic, such as glands producing an acrid secretion, that could be presumed to give it protection from predators.

These, then, are the requirements for mimicry: that the model be common, conspicuously marked, and protected by some special mechanism such as distastefulness or a sting; that the mimic be much rarer than the model, but with similar flight habits. The theory, of course, is that the predator will learn to avoid the model, and will thus avoid the mimic also. The mimic thus is condemned to rarity, because if it were common, the advantage of the conspicuous coloration would be lost both for it and the model.

This all sounds rather far-fetched as it is written down, and the laboratory biologists of the temperate zone are apt to ridicule the whole business. But again, the field naturalist in the tropics is inevitably convinced. Wasps, as one might expect, have acquired a whole host of models, especially among flies, and moths. And these resemblances are so close, and of such variety, that even the experienced naturalist is repeatedly fooled. The only way to collect certain kinds of wasp mimics is to collect all of the wasps seen, and look at them carefully before turning them loose again.

The explanation of the action of natural selection here is even more difficult than in the general case of concealing coloration because of the problem of imagining the steps whereby a moth comes to look like a wasp. What good does it do a moth to look a little bit like a wasp? I think probably the initial steps may have had a different adaptational basis. In many groups of butterflies and moths the wings have become narrow and elongate in relation to particular flight habits; in others, the normal wing covering of scales has become reduced or lost, presumably also through adaptation to some habit of the group. A combination of these two tendencies would produce an insect with a vague resemblance to a wasp, perhaps close enough to confer the beginning advantage that would be seized upon by the forces of selection.

The resemblance of an "unprotected" species to a "protected" model is often called "Batesian mimicry," after Henry Walter Bates, to distinguish it from the rather different "Muellerian mimicry," first described by Fritz Mueller, also on a basis of Brazilian experience. Mueller noticed that "protected" butterflies of quite different groups also often resembled one another closely, and pointed out that this gave all of the species concerned advantage, insofar as fewer individuals would be sacrificed in the learning process whereby predators came to associate that particular color pattern with distastefulness.

Unfortunately for the advance of science, mimicry is a phenomenon that is most conspicuous in the tropics, where research is almost entirely on an observational level. If these striking mimicking butterflies occurred in the vicinity of some of our leading universities, we would know a great deal more about the phenomenon, because it would become the subject of experiment and of the collection of statistical data. This, of course, is just one more example of the general need for work at the experimental level on evolutionary phenomena, a need that is particularly great for studies under tropical conditions.

Concealing coloration, warning coloration and mimicry serve as convenient examples of the sort of phenomena that fall under the heading of adaptations to the organic environment. The devices of flowers to ensure pollination; the devices of animals to exploit special sorts of food supplies (the neck of the giraffe, the tongue and claws of the anteater); any number of topics would have served us equally well. It should also be remembered that adaptation is as much a matter of behavior as it is of structure. The hiding habits

of animals are as much adaptations to avoid detection as is concealing coloration; rapid growth in either animals or plants may be an
adaptation to transiently favorable conditions, like the rapid growth
of desert plants after a rain. And many kinds of very special habit
adaptations have been described, like death-feigning.

ADAPTATIONS WITHIN THE ORGANISM

Huxley's third group of adaptations, those related to internal
adjustments in the organism itself, need concern us little here, since
they are primarily matters of physiology rather than natural history. Internal and external adaptations are necessarily related: the
structure and functioning of the digestive organs, for instance, must
be adapted to the food habits of the organism. But any consideration of these adaptations leads at once to the general problems of
coordination and adjustment through the nerve or endocrine systems, to matters of classical physiology.

ACCLIMATIZATION

The study of acclimatization is closely related to the study of
adaptation. I would distinguish between the two by considering
adaptation to be hereditary modification of the organism to environmental conditions, and acclimatization to be modification of
the individual in relation to particular environmental conditions.
The distinction is not fixed in ordinary usage, since we speak of
the acclimatization of a new crop, for instance, through the selection of strains suited to the conditions of the new area of cultivation; or of the acclimatization of a man to a high altitude through
long residence. The two phenomena, though, are distinct.

The effects of acclimatization and adaptation may be very
similar. A so-called "white" man in the tropics becomes dark
through the process of tanning, of the accumulation of melanin in
his skin. Native races of the same areas inherently have a larger
accumulation of melanin, a sort of permanent tanning. The pig-

mentation serves the same end in each case, through protecting underlying tissues from damaging light effects. But in the case of acclimatization, there is no demonstrable cumulative effect from one generation to another. Caucasians in Africa or Negroes in Nova Scotia continue, generation after generation, to start life with the same pigmentation distribution as their ancestors. The racial differences presumably came about through some selective agency favoring survival of lighter or darker individuals in the different environments, this selective agency working on the range of variation available in the germ plasm.

The study of acclimatization reveals the extent to which organisms with the same germ plasm, the same potentialities, may be modified by varying environmental conditions. A group of workers in California led by Jens Clausen have made a particularly detailed and interesting study of acclimatization in plants at different altitudes. To be sure that the differences observed under different conditions were purely environmental effects, they have worked mostly with single clumps of a given plant, derived from a single seed, split into several parts and transplanted to different altitudes. If these transplants are compared with indigenous plants of the same species growing at a given altitude, the extent to which the characters of the local plants are due to different hereditary traits, rather than to environmental effects, can be determined.

Similar studies have been carried out in Peru with animals, by Carlos Monge and his associates. Monge has been primarily interested in studying the effects of high altitudes on human physiology. His general conclusion, I think, is that the high altitude inhabitants of the Andes represent a rather special human group formed partly by adaptation through selection over some thousands of years of people particularly resistant to the deleterious altitude effects, and partly through a process of acclimatization, of conditioning, of each individual. Certainly residents of the high altitudes cannot with impunity go to live for long periods at low altitudes, and vice versa;

but through long residence, an individual may achieve acclimatization. Animals (including man) moved from the lowlands to the highlands, for instance, commonly show sterility, ceasing to produce viable sperm. But this sterility effect may, in time, disappear in the case of a given individual.

THE INHERITANCE OF ACQUIRED CHARACTERS

Discussion of acclimatization versus adaptation, with the meaning of the words used here, inevitably leads to the question of the possible influence of environmental factors in modifying the hereditary make-up of an organism—the question of the inheritance of acquired characters. The tanning of the Caucasian is so similar to the fixed pigmentation of the Negro, that one inevitably wonders whether the fixed pigment has not, somehow, been acquired through living over enough generations in the environment where tanning occurs.

The theory that acquired characters, like a heavy tan, can influence the germ plasm, and thus become fixed, inherited characters, has come to be called "Lamarckism," after the great French naturalist, Jean Baptiste de Monet, the Chevalier de Lamarck. "Lamarckism" has somehow got involved in politics, and currently it is damned as heresy in the United States, and exalted as orthodoxy, with the stamp of official governmental approval, in Russia. It is supposed to be the opposite of "Mendelism" which is orthodox in the United States and heresy in Russia.

This, it seems to me, is rather hard on the Chevalier de Lamarck, who had difficulties enough during his lifetime. Born in 1744, he started his career as a foot soldier; was made a lieutenant when he held together the tattered remains of his company after all of the officers had been killed. Quitting the army, he supported himself as a hack writer in the Latin Quarter of Paris for fifteen years until he was able, through Buffon's influence, to get a minor botanical job. During the revolution, the National Convention established two professorships in zoology, for which they could find no zoologists; finally Lamarck was given one of these, the other going to the mineralogist, Geoffroy Saint-Hilaire. Lamarck, at the age of fifty, with no formal scientific training, thus took up a subject in which he had no previous experience, and left on it the lasting impress of his personality.

He revised the classification of the invertebrates, using the structural scheme that still forms the basis of the classification of these organisms. He was convinced that all of the diverse classes of organisms had had a common origin, and traced out a scheme of evolution fifty years before Darwin. He had an abundance of ideas, which he developed at length in his writings. Some of these ideas seem to us, even in the context of their time, absurd; others show a remarkable and penetrating intelligence. His viewpoint was always mechanistic, and he sought to explain the direction of evolution through the action of the environment on the organism in ways that may seem childish to a contemporary biologist. But also I have the impression at times that a few of my colleagues feel that it was Lamarck's fault that, in 1809, he did not have the experimental results of 1940 at hand as a basis for his ideas.

For biologists have been defeated in all of their attempts to find a mechanism whereby the tan of acclimatization can, through direct influence on the germ plasm, be converted into the dark skin of the Negro adapted to the tropical sun. It looks as though it might happen that way, but apparently it doesn't. The evidence has become so overwhelming that it is converted into a dogma which every schoolboy must learn, "acquired characters cannot be inherited." Lamarck did not stumble onto an explanation of the mechanism of evolution that we accept today; Darwin's later and more disciplined thoughts on the same subject have stood better the test of time. The geneticists have given us a whole set of beautiful explanations of the mechanism of heredity, which should serve us solidly in explaining the mechanism of evolution.

CHAPTER XV

The Mechanism of Evolution

MOST books on evolution take up a lot of space with the review of the evidence that some process of evolution has taken place. There is no more question of this among contemporary scientists than there is of the relative movements of the planets within the solar system—once also a hotly disputed point. The situation is quite different, however, when we come to the problem of the mechanism of evolution, the problem of the nature of the process itself, and of the forces that govern its speed and direction. Here there is no universal agreement, only a collection of theories and hypotheses tying together a relatively inadequate assortment of facts and observations.

I think there is little need to review again the evidence that life as we see it is the result of some evolutionary process. All of biology bears witness to the interrelations of living things, to the development of relatively complex and specialized organisms from more simple and generalized forms, to the historical continuity of the life process. Classification is the arrangement of living forms into cousinly categories, showing the nearness or distantness of the relationship, and based on the assumption of eventual common ancestry. The traits of genera, families, orders, classes, fit into such an assumption and become absurd if tested against the only other proposed assumption—that each kind of organism has resulted from special creation by an anthropomorphic god. Anatomy and physiology make sense only if their materials are considered to be the results of evolutionary processes. The historical record of paleontology, fragmentary as it is, shows an orderly sequence in time for the appearance of the great divisions of the animal and plant kingdoms, arranged as one would expect, in a series of increasing complexity and specialization.

DARWIN AND EVOLUTION

Darwin did not discover evolution: he contributed a theory for a possible mechanism of evolution. His theory of natural selection, arrived at independently and at the same time by Alfred Russel Wallace, was so ingenious and plausible, and his marshalling of the general evidence for an evolutionary process was so masterly, that he first made evolution an integral part of the mental equipment of educated mankind. In this he was surely aided by the ripeness of the time, by the fact that biology had advanced to the point where the idea of evolution was bound to explode into the general consciousness. It is hard, otherwise, to account for the failure of Darwin's predecessors, of men like Erasmus Darwin and Lamarck, to stir contemporary interest.

I am not trying, by this argument, to belittle Darwin in any way. To me, he is one of the greatest men who ever lived, comparable in stature with Newton and Galileo in the sciences, with as great a share in the shaping of human events as any Caesar, Napoleon, Washington, or Lincoln. After all, the man, in no case, can be considered apart from the times. It is as useless to study Darwin's genius apart from the intellectual climate in which he lived, as it is to study Lincoln's apart from the political climate. To what extent the course of human events is shaped by the personalities of such men, and to what extent such men are merely the products of the ripeness of their times, is a matter for fruitful debate among the historians. The men still stand out as great figures, for study and emulation according to the varying tendencies of our tastes and our ambitions.

But Darwin did not discover evolution. He gave it common currency. To trace the history of the idea one must go back, as in so many things, to the Greeks, making wide excursions in science and philosophy. Darwin himself provided material for this in the "Historical Sketch" appended to later editions of the *Origin of Species,* and Henry Fairfield Osborn has given a more complete account in his book *From the Greeks to Darwin.*

The pre-Darwin literature is a drop compared with the bucket of post-Darwin literature. I like to collect books, and I once had the idea of forming a library on evolution, but it didn't take me long to discover the hopelessness of the project. Now, when I browse in the dealers' shops, I keep a firm grip on my pocketbook and a clear picture, in my mind, of the limitations of the shelf space at home.

If the project of forming a library on evolution was foolish, the project of giving a sketch of the subject in this chapter is probably to be classed as plain crazy. I haven't read half of the books on my own shelves, and I haven't understood a good proportion of those I have read. The project is, furthermore, dangerous, because all of my colleagues will read this book (not for information, but to find what mistakes I have made—I do the same thing with their books) and none will agree with the emphasis or the point of view, and they will be volubly dismayed at the topics I have neglected. I am caught, though, because it would be unthinkable to write a book on natural history without a chapter on evolution.

The vastness and the confusion of the subject is a reflection of our half-knowledge. If we know only a little bit about something, our knowledge can be summarized readily; but as we learn more the simplicities disappear, and summary and generalization become progressively more difficult. When, though, we have learned enough so that we begin to feel that we have some comprehension of the subject, summary again becomes easier because it can be made in terms of the general principles that we have found. The subject of evolution is still remote from this simplifying stage.

THE MECHANISM OF HEREDITY

We are dealing, really, with a whole complex of problems, for which evolution serves as little more than a common focus. There is the problem of the mechanism of heredity, the subject of the contemporary science of genetics, which was touched on in Chapter V. This has two aspects: it must explain how similarities can be transmitted through the generations; and how dissimilarities, variation, may arise. Darwin constantly regretted that so little was known about this process; but the ignorance did not prevent him from building up a very plausible theory of evolution. He knew that similarities were transmitted, and that variation did exist to furnish the raw material upon which evolutionary forces could act.

Since Darwin's day, we have learned a great deal, though I sometimes think that our vocabulary has grown faster than our information. We have located the control of heredity in something called genes which occur in the chromosomes of the cell nucleus; and we know that the potentialities of a developing organism remain the same as those of its ancestors unless something happens to change the physical and chemical patterns within or among these genes. We have learned something of the probability of such changes occurring, and we know that the rate of change (frequency of mutation) can be influenced by various things, such as X-rays. Most of the variation that we have observed is harmful to the organism and, under natural conditions, promptly eliminated.

I don't think anyone has been able to visualize how these molecular genes govern the form and function of the organism. We haven't learned how the direction of the stimulus to variation may be governed. Some people would maintain that we have learned that this variation is completely random, ungoverned, a matter of chance;

but agreement among biologists isn't general enough for this to be taken as a fact.

In short, I often feel that we have essentially done little more than restate the problem, as it existed in Darwin's day, in clearer terms: which, goodness knows, is a great advance. But we can dissect out this problem of the mechanism of heredity, leave it to one side, and find that we are still left with a hardly diminished carcass of evolutionary problems.

THE ORIGIN OF SPECIES

Darwin's great book on evolution was called *On the Origin of Species by means of Natural Selection*. It deals, however, with the entire body of biological evolution, with the progressive change, diversification and multiplication of kinds of organisms. The "origin of species" for Darwin could cover all of evolution, because he was embarked on the explanation of a process that could replace the assumption that each species of animal or plant represented a separate act of divine creation. In substituting an evolutionary process for divine creation, he had proposed a new theory for the origin of species.

Again we are now able to restate the problem. The origin of species means something quite different to the contemporary biologist from what it meant to Darwin: it is a narrower, simpler, clearer problem. Which must be chalked up as progress, even though the narrower statement of the problem cannot be tagged with a definite "solution."

We reached the definition, early in this book, of a species as a "reproductively isolated population." The problem of the origin of species, then, becomes the problem of the origin of reproductively isolated populations. How does a single population, a mutually interbreeding aggregation, come to divide into two populations, living side by side but no longer sharing in the reproductive process, each following now its independent course of evolution? It is a very special sort of problem, this fission of populations, which can be dissected away from our evolutionary carcass and given a separate label, as "the mechanism of speciation."

The taxonomists, involved with the problems of the recognition and classification of species, have made the mechanism of speciation their preferred field of study. They had the field all to themselves for a while. The early geneticists, preoccupied with the description of the mechanism of heredity, of the methods of transference of hereditary characters and of the ways in which new characters might start, did not realize that there was also a problem of how the new characters would get distributed in a separate population, so as to form a new species.

Presently, however, the geneticists did come to recognize that this matter of populations was a problem for them as well as for the taxonomists. After a certain amount of bickering, in which poly-syllabic words were recklessly coined on all sides, the two kinds of scientists settled down into an almost symbiotic relationship. The geneticists used the biggest words, and pretty well clubbed the taxonomists into submission; but the taxonomists, nursing their vocabularies, have made a rapid recovery. If the geneticists and taxonomists, with their newly won friendship, can snare a few physiologists into the alliance, progress in understanding as well as in vocabulary may be achieved.

The obvious first step, in the study of the mechanism of speciation, is to try to find out how similar species keep in sexual isolation from one another. This problem has received a great deal of attention, though its possibilities for study are far from exhausted. Often similar populations simply don't come in contact: they may inhabit different geographical regions, or different habitats within a geographical region, or reach sexual maturity at different times of year. When the two populations do reach sexual maturity in the same place and at the same time, the barrier most frequently involves behavior—the two species have different mating habits.

Sometimes laboratory experiments demonstrate that crossing can occur, with fertile hybrid offspring, though the behavior patterns prevent such crossing in nature (if they didn't, two such species would rapidly fuse into one). More often, with laboratory experimentation, various kinds of sterility barriers are found. Sperm may not reach the eggs in an animal, or the pollen tube may fail to grow in plants. If fertilization does take place, the zygote may fail to start developing, or the embryo may die in early stages, or the hybrid organism may grow to maturity but be itself sterile, as in the classic example of the mule.

The barriers that actually exist between recognized species thus show considerable variety. The next step is to examine populations that seem to be something less than species, to try to find partially isolated populations, and thus gain some understanding of the process by which isolation may be achieved. This is where the taxonomists have been particularly busy, searching for subspecies. And their greatest success has been in describing various stages in speciation in connection with geographical isolation.

GEOGRAPHICAL SPECIATION

Students of the vertebrates and of the better known insect groups (such as the butterflies) share a general conviction that speciation has occurred primarily through geographical isolation. The fact that closely related species are apt to have adjacent, but separate, geographical ranges was recognized in the writings of both Lamarck and Darwin. The phenomenon was subject to particularly intensive study in the early nineteen hundreds, when taxonomists first began to change their naming system to include "geographical subspecies."

The situation is particularly clear on islands. Each island of an archipelago may be inhabited by a slightly different form of the same species. The populations of such islands are isolated by the sea barrier, which may be more or less effective, depending on the habits of the animals, the width of the sea barrier, and so forth. Where the isolation is complete, each island population can follow an independent course of evolution which, with time, will lead to increasing divergence in appearance. Every gradation can be found on such islands, from populations that are identical as far as can be determined, through populations that look alike but that on statistical analysis are found to be slightly divergent, through populations that seem to be representative "varieties" or subspecies, to populations that are different enough so that all taxonomists are agreed in calling the inhabitants of each island a separate species. When this stage of divergence has been reached, it is easy enough to imagine some agency, perhaps a geological change uniting two islands, or perhaps some accident of dispersal, bringing the two species together again. There would then be two coexisting species, two populations where there was one before.

Geographical varieties, or subspecies, are clearest in islands, on mountains, or in mountain valleys, because the nature of the terrain gives clear boundaries and definite barriers; but geographical variation is a very general phenomenon by no means limited to such discontinuous habitats. Almost all vertebrates break up into recognizably distinct geographical forms where a given species spreads over a large continental area, and it is apparent that evolution may be proceeding at different rates and in different directions in various parts of the range. Intermediate areas, in such a continental distribution, are apt to be inhabited by intermediate forms.

Sometimes the variation is a fairly continuous gradient, called a "cline" by Huxley: individuals of a given species may gradually become smaller and darker from north to south over the range of the species, for instance. Sometimes the variation is partially discontinuous, so that a series of recognizable subspecies may be connected by rather narrow transition areas where the inhabitants have intermediate characters.

It is easy enough again in such a situation to imagine evolution

reaching a point in one part of the range of a species at which the
individuals would no longer fuse with those in another part of the
range; and reshuffling of the distribution would then give an increase
in the number of species.

This situation provides the taxonomists with inexhaustible
material for study. Opinion as to the status of populations in different parts of the range of a given kind of animal is almost necessarily
subjective, so that authorities can argue endlessly as to whether a
particular kind of field mouse in Ohio is a subspecies of the similar
mouse in Missouri, or whether they are two species, or whether they
are not different enough to warrant recognition with different names.
It is not a sterile subject for argument, because it is all bound up
with the problem of the mechanism of speciation through geographical isolation. Many students of the vertebrates argue that it is only
through geographical isolation that new species can arise. This is
reflected in their system of names, for the only "varieties" deemed
worthy of separate Latin names are those that can be defined
geographically; they, and only they, are called "subspecies."

NON-GEOGRAPHICAL ISOLATING MECHANISMS

This universality of geographical variation holds in the vertebrates, and in some of the invertebrates like the butterflies; but
any general survey of the problems of speciation leads to the conviction that there must be other kinds of isolating mechanisms besides the geographical. Among plants and among many invertebrate
groups geographical variation is not nearly as universal or as clear
as it is among vertebrates. Even among vertebrates there are some
situations where it is difficult to imagine the operation of geographical isolation: among the fish of Lake Tanganyika, for instance,
where the various ecological niches are inhabited by a series of related species that have apparently evolved differing adaptations to
the different available situations within the lake.

Students of evolution have exercised a great deal of ingenuity in trying to find other types of isolating mechanisms, but none is as clear or as well documented as the geographical. It is hard to imagine a barrier to interbreeding arising directly through mutation in a homogeneous population, because mutations occur in single individuals, and if the individual thus suddenly leaped to species status, there would be no companion with which to breed. It seems as though the barriers to breeding must arise gradually between populations that are temporarily or permanently divided by some other factor.

One can imagine an ecological barrier: one part of a population becoming isolated through a change in habitat. This would be particularly easy in the case of animals with fixed food habits. An insect species, for instance, that lived always in close association with one particular plant, might become divided into two populations through some individuals adopting a different plant host. In the case of parasites, a part of the population that got switched to some new host might be just as effectively isolated from the population in the old host as animals inhabiting two different islands. Or one can imagine some shift in mating season in a population with one generation a year—if part of the population reached sexual maturity in May, and another part in July, the two might be as effectively isolated as though by geographical barriers.

The last word has not been said on this problem of the mechanism of speciation, though a great many words have been dedicated to the subject. The books by Ernst Mayr (*Systematics and the Origin of Species*) and by Julian Huxley (*Evolution: the Modern Synthesis*) provide an excellent introduction to further exploration.

After we have dissected off the mechanism of heredity and the mechanism of speciation, we are still left with a hardly diminished carcass, and further dissection becomes increasingly difficult.

THE EVOLUTION OF MAJOR ORGANIC TYPES

Richard Goldschmidt (who lives in California) has maintained stoutly, in the face of almost universal disagreement from other biologists, that there are two very different kinds of evolution, micro-evolution and macro-evolution. The taxonomists with their subspecies and the geneticists with their large accumulations of slight mutations, are studying micro-evolution, which, he thinks, is essentially different from the large discontinuities that separate genera, families and so forth. I have never understood his arguments, so I can't give a lucid explanation of them here. Goldschmidt's terminology provides, though, a means of breaking up the remaining evolutionary carcass, especially if we add "mega-evolution" (coined by the paleontologist, George Gaylord Simpson) to cover the origin of the basically divergent phyletic stems like echinoderms, insects, fish and birds.

The accumulation of minute differences leading to constantly greater divergence between populations and groups of populations can surely account for a great deal of evolution. Among living forms we find every degree of sharpness of difference from scarcely distinguishable subspecies, through sharply different subspecies and species, to clear-cut groups of species, subgenera and genera. Similarly, in the paleontological record, gradual transitions are often found in successive deposits showing the history of changes between forms that would be ranked as different species, genera, or as members of higher categories.

But transition forms between the highest categories—orders, classes, phyla—are exceedingly scarce, either as living animals or as fossils. These highest categories represent Simpson's mega-evolution, and he has examined the possible explanation of the discontinuities at some length in his book (*Tempo and Mode in Evolution*). He comes to the conclusion that it is not necessary to postulate that these groups evolved by saltation, by jumps, despite the universal gaps in the record. He considers it probable that the intermediate stages would be represented by small populations of (relatively) small-sized animals, undergoing evolution for some reason or other at an unusually rapid rate, in the course of adaptation to new ecological situations. The chances of recovery of fossils might thus be nil.

At the moment both macro-evolution and mega-evolution are outside of the reach of our experimental control. We can study micro-evolution fairly readily under laboratory conditions. It is debatable whether anyone has produced a "new species" by laboratory procedures (a lot depends on the interpretation of the definition of species); but the whole process seems not to be out of the range of laboratory analysis, and the apparently essential steps can be tested experimentally. Under these circumstances, the proper method of science is to attempt to explain the unknown phenomena, macro- and mega-evolution in this case, in terms of the known, and to postulate a different process only if the known process is obviously inadequate. Simpson has given a plausible exposition of how the unknown can be explained in terms of the known, and the burden of proof is on anyone who tries to formulate an alternative.

This in no sense means that we should have any feeling of confidence in our present explanation. The history of science is a repeated story of adequate explanations becoming inadequate as knowledge develops, and we can thus almost be sure that our present explanations, if not definitely wrong, will at least prove to have been inadequate. This does not interfere with the intellectual service that they do us now; and it is also a basic method of science that an inadequate explanation cannot be discarded until a more adequate one is at hand. The only test of adequacy is utility, which is demonstrated by general acceptance. The only widely accepted theory of a process of evolution among biologists today is through the slow accumulation of small differences, the observable process of micro-evolution.

After all of this, I still will suggest that the mutations, the variations, that lead to striking new directions in evolution, to the development of new orders, classes and phyla, may sometimes be of a different sort from the commonly observed minute variations. A change in the onset of sexual maturity in relation to the development of the organism, as in the case of neoteny described in Chapter VI, might result in an abrupt shift in evolutionary direction. The suggestion has been made, for instance, that the vertebrates (chordates) may have had their origin in this way from larval echinoderms. Other types of variation, as drastic as neoteny, may eventually come under our observation and experimental control.

THE RATE OF EVOLUTION

The study of evolution is not made easier by the rate at which the process occurs in nature. Simpson has examined this subject in considerable detail. The horse family (Equidae) has one of the most complete of fossil records, and Simpson finds that it has taken, on an average, about 5.6 million years for the evolution of a genus.

Curiously, the length of generations seems to have little to do with the rate of evolution. Elephants, with a very long life span, have evolved at about the same rate in relation to geological time as rodents, with one or more generations a year. This, perhaps, is related to the tendency of long-lived animals to form populations of relatively few individuals, as compared with the teeming populations of animals with a short life span. Small numbers may result in a high rate of evolutionary change; large numbers in a slow rate. The effect of length of generation versus the effect of size of population might thus cancel out to equate the elephants and the rodents. (Simpson is not to be blamed for this suggestion.) The effect of population size on possible rate of evolution is, as a matter of fact, a subject of debate among the mathematicians, and perhaps best left alone until they have reached some agreement.

While the length of generations seems to be unrelated to the rate of evolution, it is clear from the paleontological record that the rate of evolution has been very different in different groups of organisms, or in the same groups of organisms at different times. Thus the crocodiles have lived through the whole stretch of Cenozoic time practically unchanged, while modern mammals went through their whole history of evolution and diversification. Yet the crocodiles were evolving rapidly enough in the Cretaceous.

It is extremely difficult to study a process by the experimental method, if that process takes millions of years. One is tempted, at times, to shrug off the whole problem of the mechanism of evolution as insoluble with our present tools because of the time scale, and because our knowledge of the environment in times past is based wholly on inference. But to be thus easily defeated is to abandon not only this area of science, but the whole of the scientific method. Erosion and mountain uplift are slow processes, but not inaccessible to study for that reason. Nor is the rate and nature of the decomposition of radioactive ores. Seas may become dry land and the land be elevated into lofty mountain chains while a genus or family of insects or mammals is undergoing its evolution. And the bit of the process that we see in the mutations of our laboratory populations or in the variations of wild specimens is like the annual sediment of a mountain stream. The study is certainly not made easier by the time scale involved, but I cannot believe that it is out of the reach of our methods of study.

THE EVOLUTIONARY DRIVE

Let's look again at our dissected carcass. We have taken out the mechanism of heredity and the mechanism of speciation, and we have divided the remainder into the micro-evolution of field observation and laboratory analysis, the macro-evolution of species and genera, still hardly under our experimental control, and the mega-evolution of major organic types, whose study is the prime concern of the paleontologists. But what makes the thing work? What is the drive, the unifying and coordinating force?

Darwin thought the answer was natural selection. The force of his book comes from the fact that he had found a unifying principle that tied together the most diverse phenomena; that seemed to furnish a drive for all of evolution, once given the initial impulse. Darwin apparently did not worry about that initial impulse, about the origin of life. He was content to leave origins to God, and to dedicate his own efforts to searching out the patterns that control development.

We still, I think, have managed little improvement on Darwin's theory. The woods are full of neo-Darwinians, anti-Darwinians, Lamarckians, Bergsonians, orthogenicists, and goodness knows what else, all with theories to sell, mostly wrapped in almost impenetrable layers of verbiage. The core, when finally arrived at, usually looks to me like metaphysics—an *élan vital,* an innate tendency of some kind beyond the reach of the experimental method. Such theories then belong to the province of the philosophers, and ought to be examined there. Whether true or not true, they are of no use to the naturalist as a naturalist, even though as a man they may soothe his human longing for universal explanations.

I can do no better in this regard than quote Simpson on the same subject: "Scientific history conclusively demonstrates that the progress of knowledge rigidly requires that no non-physical postulate ever be admitted in connection with the study of physical phenomena. We do not know what is and what is not explicable in physical terms, and the researcher who is seeking explanations must seek physical explanations only, or the two kinds can never be disentangled. Personal opinion is free in the field where this search so far has failed, but this is no proper guide in the search and no part of science."

When I read a modern attempt at a synthesis of our knowledge of evolution—Julian Huxley's is probably the most comprehensive

—I am left with a strong feeling of dissatisfaction. The various pieces, the theories, the accumulation of factual observations and experiments, are all very impressive, but they do not seem quite to fit together. It all is rather like a super-saturated solution waiting for the right idea to precipitate crystallization.

Darwin's book performed such a function for the accumulated information of 1859. Surely some penetrating intelligence will add the idea that will form a sensible pattern out of the information that we have today. I see no need for appeal to the metaphysical, even though, equally, I have no hope that the theory that helps us today will serve a hundred years from now. It is simply that in the endless chain of links of observations and experiments with conceptual schemes, we have reached the point where a new conceptual scheme is needed. When our data have thus formed a challenge, the response will surely come.

CHAPTER XVI

Natural History and Human Economy

THIS account of evolution completes the sketch of the subject matter of natural history. We have discussed the naming and cataloguing of organisms; their reproduction and development; their relations with the environment and their organization into populations and communities; and finally, their evolution, the explanation of the diversity of organic form and of its fitness or adaptation. I ended with an emphasis on ignorance, which is not a neat way to tie up a parcel of knowledge. But all biologists are more impressed by what we don't know than they are by what we know, and such an emphasis is thus a true reflection of our state of knowledge and our state of mind.

My object in this book has been to describe an area of science: using the particular field called natural history as an example, or type case of the general phenomenon of science. Such a description requires much beyond an inventory of the subject matter of the science, and I have tried, all the way through the book, to give some attention to attitudes, to historical backgrounds, and to personalities.

Three further general topics, however, surely warrant separate chapters, to round out this description of a type science. First, there is the question of the application of science, of its relation to human economy, which forms the subject of the present chapter. Then there is the question of the kind of men who create science—the scientists, their motivations, drives, quirks and limitations—which form the subject of the next chapter. Finally, there is the question of the nature of the scientific method, which may very appropriately be covered in the final chapter of the book because in a real sense it serves as a summary of all that has gone before.

SCIENCE: PURE AND IMPURE

There are two very different points of view with regard to the relations between science and human economy. By one group, science is regarded as wonderful because of the material benefits that have accrued from its development. Chemistry should be cultivated because it has given us plastic fountain pens, and may presently give us synthetic eggs. Physics deserves our wholehearted support because it has produced the radio, the electric motor and the atomic bomb.

By the other group, viewing science and human economy from the second point of view, we are warned about the danger of keeping our eyes too closely on the applications of science. This group points out that the applications of science depend on the development of pure, or theoretical knowledge. Thus the greatest marvels of applied science involve phenomena of electricity; yet electrodynamics, which made these marvels possible, was developed by men with no interest in the practical possibilities of their work— men like Clerk Maxwell and Willard Gibbs. Examples of material progress resulting unforeseen from progress in the theory of science could be multiplied indefinitely, but they have surely been sufficiently advertised to underline the moral.

Actually, I suspect it is impossible to disentangle "pure" science and "applied" science. A more valid distinction can be made between science and technology, in our contemporary world, if science is considered to be composed of experiments and observations that lead to the formation of conceptual schemes, while technology is considered to involve experiments and observations directed toward the solution of particular problems without relevance to new conceptual schemes. Even this distinction seems less clear as we go backward in time.

The English biologist, Lancelot Hogben, has argued persuasively in his book *Mathematics for the Million* that the development of science has depended on a close association between the thinker and the artisan, on the development of human economic and material needs. The sterility of Greek scientific thought, its failure to grow beyond its philosophical origins, according to Hogben, was caused by the complete separation of the thinker and the practical man in the organization of Greek society; by the fact that in a slave society, there, was no need or possibility for the educated man to have physical experience with the everyday problems of agriculture, mining or industry.

Hogben and other proponents of this view have shown how modern mathematics had its origins in the problems of calculating trajectory in artillery fire. Similarly, the theory of the vacuum and other early physical discoveries clearly arose from such practical problems as the designing of pumps. Early natural history had its origins in medicine, in the curiosity of physicians concerning the functioning of the body, in the searchings of the herbalists for new plants that might be useful in curing disease.

The great figures of science have often been very practical men, interested in practical problems. In biology, aside from the physicians, we have men like Anthony van Leeuwenhoek who was a mechanic (as well as a linen draper), grinding his own lenses, as much interested in the manufacturing problem as in the tiny animals that he discovered with his product. Louis Pasteur is the archetype of the practical man pushing ahead the theory of science, because Pasteur started always with a definite, practical problem—

the souring of wine, the devastation of the silkworm industry—even though he might end with a grand and fruitful general theory.

The logical division, then, is not between pure and applied science, but between science and technology: between observations and experiments on materials of whatever sort that lead to the formulation of fruitful conceptual schemes; and observations and experiments that are complete in themselves, that have no bearing on the discovery of new relations among natural events. The difference is perhaps one of personality, of the mental outlook of the investigator. I think everyone would agree that Gregor Mendel was a scientist, and Luther Burbank a technician. Yet both worked on garden plants. The one, with his few simple experiments, opened up a whole new field of thought; while the other gave us merely bigger raspberries and brighter roses. Mendel used his experiments for the development of a conceptual scheme, while Burbank was interested only in the physical product. The genius of Pasteur lay in combining the two: he got his practical results, but along with these, he got new and fruitful conceptual schemes.

THE APPLICATIONS OF NATURAL HISTORY

In the case of natural history, a consideration of the inter-relationships of the science with human economy gets us involved again with the verbal problems of the divisions of the biological sciences. "Applied natural history" is a phrase that carries little meaning. It brings to mind at most a forest ranger, or a member of the Audubon Society explaining the value of robins because of their worm-eating habits. But if we stop to think, it is a phrase that should have very wide implications indeed.

I defined natural history as biological investigation at the level of the individual organism: the study of the relations of organisms among themselves and with the physical environment, and of their organization into populations and communities. When we stop to think about the ways in which studies of this type impinge upon our practical affairs, we find that we are dealing with areas of investigation that have been separated off under a variety of names: agriculture, forestry, conservation, epidemiology, climatology, and goodness knows what else.

It is interesting to look at the broad fields of agriculture and medicine as areas of application of biology. Both deal with the practical manipulation of living things and processes: clearly they are biological; equally clearly, they are applied science. Their relationship to "pure" biology, however, is very different from the relationship, say, of electrical engineering to physics. Both agriculture and medicine existed as technologies long before science in the narrow sense began to emerge as a recognizable human activity. The technologies gave rise to the science. In physics and chemistry, on the other hand, the sciences have given rise to the contemporary technologies. This results in a different relationship between theory and practice in the biological and in the physicochemical sciences.

AGRICULTURE AND NATURAL HISTORY

Agriculture is a vast subject—particularly if we use the word in its most inclusive sense, covering all activities directed toward the utilization in human economy of other organisms, whether animal or plant. Yet a great part of agriculture, whether concerned with the cultivation of plants or the husbanding of animals, falls at the level of natural history. The problems of agriculture are the problems of the interrelations of organisms with each other and with the physical environment, from the particular point of view of gaining information that will enable man to manipulate the organisms to his own maximum advantage.

Natural history, as a theoretical science, owes a great debt to agriculture. This can be realized very clearly by a review, for instance, of the writings of Charles Darwin. He depends all through the *Origin of Species* on the accumulated observations of the practical men, the farmers, the animal and plant breeders. In the two volumes on the *Variation of Animals and Plants under Domestication,* he summarized a mass of practical agricultural knowledge into an orderly contribution to natural science.

Much of our knowledge of the food requirements of plants and animals has sprung from the needs of agriculture. The intensive study of the nitrogen cycle has been spurred on by the necessity of improving crop yields. Genetics is as thoroughly entwined with the practice of agriculture as it is with the theory of evolution. Animal and plant behavior, climatic adaptations, soil formation and modification, are equally the province of agriculturists and naturalists, and up until now the naturalists have learned more from the agriculturists than vice versa.

There is, however, a growing realization that the practical agriculturist can also learn a great deal from the academic studies of the naturalist. The realization has come first in problems of land management, of forest conservation, where the applicability of the ecological studies of natural communities is clear. The exploitation of fisheries, game and other wildlife resources, has also come to be based firmly on ecological knowledge.

Biological studies are recognized as a necessary basis for training for agriculture in the strict sense—the management of field crops —and many modern farming practices involve the application of knowledge gained by the naturalists. The debt to genetics has already been mentioned. The control of pests, of insects, fungi, bacteria and other parasites of cultivated plants, has come to depend more and more on studies of the life histories, adaptations and habits of these parasites—on studies at the level of classical natural history.

The control of pests through the modification of environmental factors, rather than through the direct application of poisons, has come to be known as "biological control," and is the subject of a voluminous literature. Man, in transporting crops from one part of the world to another, has also transported crop pests; and these pests, introduced into a new environment, have sometimes caused great damage. Perhaps the most successful of biological control measures has been the introduction of natural enemies of these introduced pests. The ravages of the exotic insect pest in a new country are generally due to its escape from the parasites and predators that keep its population in balance in its native home. The logical course, then, is to search for and introduce these parasites and predators. The method is not infallible, but the results in some instances have been spectacular.

MEDICINE AND NATURAL HISTORY

The relation of medicine to biology is even closer than that of agriculture, but this relation involves the natural history sciences less directly. The biological basis of medicine lies rather in physiology and anatomy, in the functioning of the parts of the individual. There is one great area of overlap, however, between modern medicine and natural history—the study of the phenomena of parasitism.

This study is divided, rather vaguely, between two sciences, parasitology and epidemiology. The parasitologists study the life histories of parasites using the conventional methods of the naturalists—describing form and function in different stages, host relations, methods of dissemination, geographical distribution, and so forth. The epidemiologists are concerned primarily with the dynamics of the relationships between parasites and hosts, and thus are studying the same sort of phenomena as the naturalist when he is concerned with the dynamics of populations.

Epidemiology got its start through the practice of keeping statistical records of the incidence of disease, of epidemics. An epidemic is an unusual outbreak of a disease. This contrasts with endemic conditions, in which a disease is constantly present, with an incidence reflecting a more or less balanced relation between parasite and host. It soon became obvious that epidemic conditions could not be interpreted except through the study of endemic conditions, giving rise to the verbal contradiction of epidemiologists studying endemic diseases.

The contrast between endemic and epidemic conditions is sometimes geographical. Yellow fever, for instance, was formerly endemic in many parts of the American tropics, extending to the United States at irregular intervals through the accidental introduction of the virus, where it caused fearful epidemics. Climatic conditions did not permit the virus to maintain itself indefinitely in American cities, so that the disease could never become endemic in those latitudes.

Or the contrast between epidemic and endemic conditions may be due to environmental changes. The disturbed conditions of war and post-war periods produce many environmental changes that permit diseases, always present at a low endemic level, to flare up into epidemic proportions, so that war and disease have formed a partnership that has persisted all through human history. An unusual climatic condition, a particularly cold winter, or warm summer, or unusual rains, may, through some modification of the environmental relations of man and parasite, produce an epidemic.

The epidemiologist, then, in trying to describe the incidence of disease, is involved directly with the problems of ecology, of natural history. The incidence of the disease, he has found, depends on the relationships between the parasite and the host, and on the relationships of each of these to the factors of the physical environment. Where the parasite causing the human disease has an intermediate host, or vector, as with yellow fever and malaria, the epidemiologist becomes even more obviously an ecologist, studying the environmental relations of the vector.

The philological problem presented by the word "epidemiology" has long amused me. The epidemiologists adapted themselves without difficulty to the fact that they were studying endemic diseases. But when the study of human diseases involved them with other animals, they got tangled with the etymology of their label.

"Epidemic" stems from *demos*, people; so that while rabies might become epidemic among people, it has to be epizootic (*zoön*, animal) among dogs. If a man is studying the incidence of measles in children, he is safely housed under epidemiology; but if he becomes interested in the incidence of tularemia in rabbits, he must pack up and move over to epizoology. The plant people, not to be outdone, have erected a compartment for plant diseases, called epiphytology (*phyton*, plant). Which is why I think they would all do better to stick to natural history. Or, if we must have a word for the special case of the study of the natural history of disease, stretch epidemiology to cover tularemia in rabbits and blight on potatoes as well as measles in man. The Greek scholars may be pained, but the dictionary people will be saved a lot of trouble.

CONSERVATION AND NATURAL HISTORY

Forestry, wildlife management, fisheries management and conservation studies in general differ historically from agriculture and medicine in that they are younger than the theoretical biological sciences. They come closer to representing applied natural history in the sense that electrical engineering represents applied physics. The analogy with electrical engineering ends, however, with this historical comparison, because the conservation sciences are far from representing an organized body of applied knowledge stemming from a clear-cut accumulation of theory.

The conservation sciences are diffuse, in part contradictory, in part mutually exclusive, with an uncertain status in the hierarchy of professions. They share these characteristics with the social sciences, which also, in a paper scheme, might be considered to represent applied biology. The conservation sciences (and the social sciences) are young, and their diffuse subject matter and ill-defined principles are often considered to be an attribute of this youthful nature. This, I suspect, is too easy an explanation.

In part I think they merely reflect the vagueness and uncertainty of their parent biological sciences—which cannot be considered "young" relative to the physico-chemical sciences. And in part, also, they reflect the complications of the emotional attitudes that color their subject matter. Argument over the arrangement of an electrical circuit does not affect the endocrine balance of either scientists or politicians; but the maintenance of forests, the preservation of duck breeding grounds, the stocking of trout streams, the extermination of coyotes, involve so many vested interests, so many childhood prejudices, that it is difficult to be objective, precise, "scientific," in reviewing the data and determining a plan of action.

The conservation sciences have developed as a result of many different pressures, and these pressures do not make the development of a unified theory or plan any easier. Man has been remaking the surface of the earth for a couple of thousand years now, but it is only in the last two or three score years that he has come to realize the implications and possibly disastrous consequences of this activity. The pressure to do something at once, to make up in ten years for a thousand years of neglect, is consequently enormous. One group insists that we must do something about the forests, another that we must do something about soils, another that we must do something about watersheds, another that we must do something about ducks or black bass or deer. The supporters of the various causes may be equally sincere and equally insistent and yet advocate diametrically opposed means of attaining their respective objectives.

Often we simply haven't got the knowledge necessary for a decision. It may seem advisable, for instance, to create a marsh to foster duck breeding, or water level maintenance; but public health people may argue that the marsh represents a malaria hazard, agricultural people that it involves a dangerous loss of needed cultivation. The duck people, the water level people, the public health people, the agricultural people, should all be applying different parts of the general science of biology (of natural history) to this single specific question. Yet their proposals may be completely different, and when they sit down together to work out a plan of action, they are apt simply to get mad and call each other names.

The conservation sciences have shown a great development in recent years in the United States. Several of our major universities have schools of forestry with large staffs and many students; the wildlife management groups have shown a comparable growth. Many aspects of agriculture and public health are concerned with the same sort of problems. The need, I think, is not so much to foster more growth in any of these specific fields, as it is to foster cross connections of ideas and knowledge. And this, I suspect, can best be done by the development of the academic natural history sciences which should form the common background of all of these diverse applications.

THE QUESTION OF VALUES

I have tried, then, to indicate the relation between natural history and human economy through such fields as agriculture, medicine and conservation. There is little use here in trying to compile a list of specific gains to civilization through such applications of science, since the gains, the "utility of science," have been very adequately stressed by everyone who has undertaken the task of "popularizing" science.

The gains, I think, are real. As individuals, we like comfort and we dislike pain, and science has helped us increase comfort and lessen pain. When a child has got a severe bronchitis that might at any moment turn into pneumonia we don't question that penicillin represents something that can definitely be called progress. Nowadays we are free from the fear of yellow fever that haunted our ancestors when they ventured into the tropics. We know that our faces and the faces of our children will not be scarred by smallpox.

The gains are real, but I doubt whether they are as great as our popular science writers would lead us to believe. What is the progress represented by the radio if, when we turn it on, we hear only the same ranting demagoguery that, a century ago, we could have heard by attending the town meeting? The airplane takes us from New York to Paris in a day: but we must spend a week collecting the visas, certificates, permits, the endless pieces of paper necessary for the trip. In 1910 it took ten days or so to travel to Paris, but the bureaucrats had not yet thought up their complicated methods of utilizing paper or of wearing out the nervous reserves of the prospective traveller. And even if we "save" nine days in getting to Paris, progress cannot be measured except in terms of how these "saved" days will be "spent."

The farmer grows four times as much corn from an acre using hybrid seed—but there are four times as many people in the world to be fed. I can view my baby's bronchitis calmly because I know that at any time of day or night, all of the resources of modern medicine can be brought to bear on the progress of the infection. But what does that mean to the father in a thatched hut far out in the Colombian llanos, or to a Chinese coolie, or to a Negro in a cold and dreary shanty somewhere in Alabama? We are all individuals, all humans, all of us tremendously important to ourselves. And progress in the reduction of pain and fear cannot be measured in terms of your pains and fears, or mine only: it must be measured in terms of the reduction for all of the millions of individuals that go to make up our species.

Thus, while it is easy to make out a case for the importance of the applications of science, it is also easy to question the value of these applications. Any evaluation, any measurement in terms of "progress," would of course depend on a definition of the objectives. What do we want, anyway? What is our "purpose," as individuals and as a species, in living? Without knowing this goal, I don't see how we can compute the rate at which we are travelling toward it, nor indeed have any idea as to whether we are correctly aimed. Perhaps we are merely running in circles through the woods; or perhaps we are accelerating our descent into the limbo where we shall join company with the trilobites and the dinosaurs.

The scientists generally shrug this question away. It is a problem, they say, for the philosophers, the statesmen, the theologians. They are not quite sure whose problem it is, though they are very sure that it is not theirs.

I can't agree with this. The whole concept of progress is something that was foisted on the human mind with the advent and development of science, so that it seems difficult for the scientists to escape all responsibility for measurement of its direction and rate.

And this, I think, should be the most important contribution of natural history to human economy. Man, inescapably, is an animal, a part of the biosphere, a member of the complex of organisms that carry on the living process on the surface of our planet. To get perspective on his own problems, to find the significance of his own feelings, reactions and behavior, it seems to me that man must first achieve an understanding of the living universe as a whole and of the interrelations of its parts. With this background, he stands a better chance of determining his own goal, of finding methods of achieving it, and of measuring his progress.

The great problems of mankind, in the words of Raymond Fosdick, are to determine purpose, to find values, to achieve social wisdom. Fosdick considers that science has failed to help in the finding of solutions to these problems, and that the failure is inherent in the nature of science. I think rather that we have failed to explore science from this point of view, accepting its material contributions and at the same time rejecting the implications of its facts and methods for our philosophy.

Purpose is a tricky word, one that the scientist in his daily thought avoids as far as possible. He feels that his province is to determine how things work, not why they exist. The ultimate purpose that he can find in organic nature is survival, as I pointed out in an earlier chapter: survival of the individual, of the species, of the life process itself. This seems at first rather bleak. But if we could accept its implications, we might achieve a great gain, since an amazing range of human activities have become perverted to the point where they endanger, rather than promote, survival—whether of the individual, the species, or the life process.

We have, particularly in the last few centuries, escaped from the controls that maintain balance and proportion in the biotic community. We have run wild, like a weed escaped on a new continent. We have retained the birth rate that we acquired during our pleistocene evolution through savagery, and at the same time radically altered the nature and the incidence of the factors causing death. The result is a density of population that exceeds all reason, all possibility of support, and the end is still not in sight. This is, beyond any question, the greatest problem that faces man. It is a problem that becomes obvious to any student of natural history who stops to look at the condition of his own species, and it is a problem in which the naturalist must cooperate in finding a solution. From it comes war and starvation and wastage of our inheritance: an accelerating wastage that we may never be able to replace, even though we learn to check it. The problem has been clearly stated by one naturalist, Fairfield Osborn, in his book *Our Plundered Planet*; it should be a major preoccupation of all naturalists.

The philosophers have long pondered on values and on purpose. They have protected the area with "no trespass" signs which the scientists, convinced of the superior grass of their own pasture, have respected. We must break down this fence and take away the signs. For the scientist, as a man, cannot escape values or purpose.

More is involved than breaking down a fence. It would be my hope, at least, that the scientist, in looking at values and purpose with the background of his own methods, attitudes and knowledge, might make his greatest contribution to human economy: a contribution that would outrank all his gadgets, pills and sera in the increase of comfort and the lessening of pain.

CHAPTER XVII

The Natural History of Naturalists

NATURALISTS are the causative organisms of natural history. One might, in fact, view natural history as a sort of secretion of naturalists—science as a secretion of scientists—and such a view makes clear the importance of studying naturalists (or scientists) in any attempt to gain an understanding of natural history (or science).

It seems to me an odd trait of our culture that we are very interested in personalities in general, but very little interested in the personalities of scientists, even though we all recognize that the product of their activities, science, is exercising a controlling influence on our fates. With the exception of Sinclair Lewis' *Arrowsmith,* I can't think of any successful novel that deals with the analysis of the scientific personality. Our novelists dissect at length the characters of business men, prostitutes, preachers, writers (with understandable frequency), bums, every kind of character except the scientist. Or when a scientist is brought in, he is some wooden thing, whittled out of philanthropic motives, draped in a white coat and saving humanity from something or other.

There are exceptions. I thought Marquand's *Point of No Return* included a sympathetic full-length portrait of a social scientist at work—though the social scientists may doubt the sympathetic part. I also found the marine zoologist in Steinbeck's *Cannery Row* both amusing and plausible, though again marine zoologists may regard such a portrait as libel.

A naturalist, according to my concept of natural history, would be a person who is studying the phenomena of life as shown by whole organisms: their development, their characteristics, their environmental relationships. It is sometimes said that the day of the naturalist, in this sense, is past, because no man now can hope to master the complexities of all of the specialized sciences into which natural history has been divided. But that is like saying that the day of the historian is past, because too many details of history have accumulated. It has never been possible to master all of the details of knowledge, and a man is a historian or a naturalist not because he has memorized a thousand or a hundred thousand facts, but because of his point of view.

The naturalist generally restricts himself in active research to certain kinds of organisms, such as birds or fungi or butterflies; or he may restrict his researches to certain kinds of phenomena, as with parasitology or ecology. But he remains interested in the general implications of his studies, not content to regard the facts about birds or parasitology or fungi as complete ends in themselves.

The word "naturalist" has got rather into academic disrepute in recent years, I think partly because it has come to be associated with another term, "nature lover." "Love" implies an attitude to the beloved through which its characteristics are transfused, given a special meaning to the lover. Blemishes become adornments and to question is to betray. Truth, objectivity, criticism, patient dissection, all go down the drain. The lover views the beloved through rose-tinted spectacles; and when someone else borrows the spectacles, the tint seems not rose, but a nasty pink. The scientist wants optical glass, not rose tinting.

"Nature lover" for the scientist has thus come to mean a sort of pervert. Nature loving twists the normal, the objective, the healthily reproductive; and, what is more, carries with it scientific ostracism. (The danger isn't in loving nature, but in admitting it.) The identification of naturalist with nature lover isn't quite complete as yet, since no scientist would dare to write a book with the title "A Nature Lover in Siam," though he might still use the title "A Naturalist in Siam." But very few of our upper-crust academic biologists have ventured to call themselves naturalists in recent years; they feel that it is safer to use labels derived from Greek roots.

I am, then, using the word here for people who might prefer to call themselves ornithologists, epidemiologists, entomologists, mycologists, and so on. It is a sufficiently varied assemblage so that one can well wonder what are the common characteristics of its members. What is their course of development, their behavior, their habitat, and how do they go about their business of manufacturing natural history?

THE DEVELOPMENT OF NATURALISTS

The commonest first sign of a developing naturalist is the collecting habit. The psychologists may have made a careful study of collecting instincts in man, but if so I haven't seen it, though much has been written on the matter by non-psychologists. In our civilization, a tendency to collect all sorts of things from earthworms to bottles shows itself rather early in childhood, and very often gets moulded by some series of accidents into a particular channel: the collection of toy train equipment, of pictures, of postage stamps, or of some kind of natural objects such as rocks or beetles.

The collecting habit is pretty common among both birds and mammals, as witnessed by the conventional references to magpies and pack rats, but I doubt whether it would be possible to equate the human habit with any general animal instinct. I suppose one could find examples enough of collecting in primitive cultures, but as a widespread phenomenon, either of children or adults, it strikes me as rather peculiar to our own particular culture. At least, we have systematized and popularized collections of all sorts of things to an extraordinary degree.

I have often wondered how, in childhood, this generalized collecting habit becomes directed toward natural history objects. I doubt whether deliberate parental direction is important, because there seems to be no clear tendency for children to follow the collecting habits of their parents.

I can remember in my own case starting to gather all sorts of things like rocks and beetles when I was about nine years old. There was no parental encouragement—nor discouragement either—nor any outside influence that I can remember in these early stages. By about the age of twelve, I had settled pretty definitely on butterflies, largely I think because the rocks around my home were limited to limestone, while the butterflies were varied, exciting, and fairly easy to preserve with household moth-balls.

I am sure that all of the external influences, books, encouragement and apparatus, came after I had fairly well settled on this butterfly path from some inward motive or other. I was fourteen, I remember, when these influences had become strong enough so that I decided to be scientific, caught in some net of emulation, and resolutely threw away all of my "childish" specimens, mounted haphazard on "common pins" and without "proper labels." The purge cost me a great inward struggle, still one of my most vivid memories, and must have been forced by a conflict between a love of my specimens and a love for orderliness, for having everything just exactly right according to what happened to be my current standards.

I didn't intend to write an autobiography, but I know my own history better than that of other people, and it seems to be typical of a wide variety of naturalists at the same stage of development. A very high percentage of them, at the age of fourteen or so, were collecting butterflies, whatever their future line of development; I think the proportion starting with shells, beetles, bird eggs, or similar things, would be considerably smaller.

Another group of naturalists start as hunters or trappers or kids that just like to mess around with animals (I suspect that an interest in plants is liable to have a later development). A third group start out as tinkerers and get into biology through some laboratory approach—but the tinkerers are more apt to develop into chemists and physicists.

In general, the development of personality and interests of the potential naturalist is well along before he reaches college. This development may get suppressed, diverted, or encouraged during the course of college education, but I think it is rarely initiated that late. Most commonly, it is suppressed or diverted, either by pressures to find some method of making a living, or by distaste for the academic approach.

The pressures to make a living must rule out all except the hardiest, because the road of the naturalist hardly appears to the undergraduate (or his family) as a broad highway to fame and riches. The interest may persist as a hobby when the boy has grown up to be a banker, lawyer or physician, and may flare up again as a full-scale activity quite late in life. The contributions to science of these amateur naturalists are considerable.

It is one-sided to stress the role of collecting in the development of a naturalist as much as I have stressed it here. Collecting and hunting interests gain their chief importance because of the neatness with which they serve to distinguish the budding naturalist from the budding scientists of other varieties. In any broad view of the development of scientists as a class, curiosity surely would have first place.

All children are curious and I wonder by what process this trait becomes developed in some and suppressed in others. I suspect again that schools and colleges help in the suppression insofar as they meet curiosity by giving the answers, rather than by some method that leads from narrower questions to broader questions. It is hard to satisfy the curiosity of a child, and even harder to satisfy the curiosity of a scientist, and methods that meet curiosity with satisfaction are thus not apt to foster the development of the child into the scientist. I don't advocate turning all children into professional scientists, though I think there would be advantages if all adults retained something of the questioning attitude, if their curiosity were less easily satisfied by dogma, of whatever variety.

That is why I think it would be useful if something of the attitude and method of science could be taught to all of our university students, whatever their direction of specialization. Memorizing the facts of science, the names of bones or the symbols of chemical formulae, is of no help toward this objective. The historical approach suggested by Conant would, it seems to me, be more fruitful. But it is easy for outsiders to throw stones at our educational methods; and the subject is only indirectly relevant here.

THE HABITAT OF NATURALISTS

We had better skip the thorny problem of the classification of naturalists into different varieties, and go on to the question, where and how do they live?

Naturalists have three main habitats: in universities, in government service, and in museums and research institutions. Not long ago, an overwhelming majority of naturalists were found in the universities, but those in government service may now be more numerous. This is a guess, because I have no statistics.

The university is still the stronghold of the naturalist, despite the recent invasion of other habitats. In the universities, the naturalist divides his time between teaching and research. Much university effort goes into research and sometimes I think that our great universities have forgotten that one of their functions is to teach, because they have become preoccupied with attracting research workers, famous names, to their faculties. These often grudge the time spent in preparing and delivering lectures, in counseling students; they speak of their "teaching load," and carry this load with obvious sweat and dislike.

Yet the two functions of teaching and research in science are hardly separable. Teaching by a man with no research interests, with no curiosity, with no drive to push out the frontiers of his area of knowledge, would surely become a sterile affair, a passing on of dogmatic facts. And research without teaching is equally in danger of becoming sterile. I think a desire to teach must be an important (even though hidden) part of every scientific personality. For one thing, there is no use in gaining new knowledge if it can't be imparted to someone; if not to students, to one's colleagues or to the world at large, which is also teaching. And scientists with no teaching responsibilities almost inevitably start instructing their assistants, their technicians, their wives and families, their neighbors, or anyone else who happens to be handy. Here am I, for instance, an orthodox research scientist with frustrated teaching instincts, trying to instruct anyone who can be persuaded to read this book.

The government employment of naturalists is a direct tribute to their importance in relation to the problems of human economy. In the United States both the state and national governments employ thousands of investigators in connection with their programs of agricultural research, including a very high proportion of naturalists—entomologists, botanists, bacteriologists, geneticists, ecologists and so forth. A few, especially entomologists, are employed in connection with health and medical programs, an employment that will no doubt increase as the value of the natural history point of view in this sort of work is more widely appreciated. The wildlife and conservation programs are using a constantly increasing number of naturalists of several kinds, and the naturalists in turn are demonstrating the practical importance of their work in these fields.

Museums and research institutions are often adjuncts of universities or governmental agencies, and their classification as a separate habitat may be rather artificial. In the museums the collecting instincts of the naturalists have full sway and the results, in terms of millions of specimens, are overwhelming. The stuffed birds and mounted butterflies that the visitor sees on exhibition are a minute fraction of the study material that is stored away in the research galleries.

Government scientists and university professors are apt to feel neglected and underpaid, but they enjoy both wealth and limelight in comparison with the museum curators. To be a museum curator requires either an overwhelming love of the subject, or a wealthy grandfather, or (preferably) both. Yet these museum curators, with their patient cataloguing, provide the whole structure of classification on which all of the rest of natural history depends. Maybe they should form a union and go on strike, so that all of the more fancy and spectacular scientists would realize their dependence on the curators. But this is unthinkable, since the curators work not for common fame or for money, but from some inner compulsion that we call "love of the subject." The trouble is that wealthy grandfathers are becoming scarce at a time when ever more curators are needed if the whole growth of the natural history sciences is to keep in balance.

THE BEHAVIOR OF NATURALISTS

Scientists are people and thus their behavior is essentially human behavior: but while human behavior has basic uniformities, it also has an astonishing diversity. Anthropologists tell us that behavior patterns are largely conditioned by the culture in which an individual lives. Under given circumstances, a Samoan will do one thing, a Fijian another, a Chinese something else, and so on. In very complex cultures, like that of our Western civilization, behavior may vary according to economic or social class, according to occupation, or to any of the other numerous groupings into which the society can be subdivided. An inquiry into the behavior of scientists in general, or naturalists in particular, would then turn on the extent to which it agrees with human behavior in general or with the behavior of the particular culture, and on the extent to which it diverges, characterizing this particular grouping of individuals.

Science in the narrow sense in which I have used the word in this book is a peculiarity of Western civilization. Its roots can be traced to the Graeco-Roman, the Arabic, the Indic civilizations; but its growth and flowering has occurred in the modern Western world. Indians, Chinese and Japanese have cultivated science in recent years with, in many instances, distinguished success; but it is still something that they have imported from the Western world into their contemporary culture, and the success of such transplants is somewhat uncertain.

We are safe, then, in examining the behavior of scientists in the context of Western culture. I suspect that scientists have created that culture, or at least played a dominant role in determining its direction and development. But cause and effect relations are difficult to untangle: maybe the culture created science instead. At any rate scientists have attained a moderate degree of respectability within the culture, ranking along with bankers and physicians in prestige if not in take-home pay. The prestige has increased greatly since Hiroshima—which illustrates human behavior rather than scientific behavior.

Scientists show no clear behavior peculiarities except in relation to their work. This is curious because their work involves a very peculiar element, objectivity. Scientists in the laboratory and in the field behave in an inhuman way insofar as they try to examine the universe without prejudice. This is an ideal that they do not always attain, but they at least try. It is an expressed ideal, maintained by constant vigilance and criticism. The scientists cling to their pet theories as lovingly as anyone, and the theories could readily grow to prejudices except for the vigilance of co-workers. If the facts contradict the theory, it has to be discarded, and any scientist who clings to an outworn theory because of his love for it is thereby excommunicated and forever damned.

Probably no man ever achieves complete detachment, complete objectivity; but the great success of the scientists lies in the fact that they try. The ease with which objectivity is attained seems to depend on the distance of the subject from human emotional concerns. Physicists and chemists have attained a greater objectivity than biologists, because the forces and substances that they study are less emotionally charged than the organisms studied by the biologists. Physiologists, I should guess, are more successful than naturalists; and naturalists certainly achieve objectivity much more easily than anthropologists or sociologists. Whether it is possible for man to achieve objectivity in studying himself is perhaps debatable—and to that extent, the applicability of the scientific method to human affairs is also debatable. But as the Harvard anthropologist, Clyde Kluckhohn, has pointed out, data are never "subjective" or "objective": it is the way of looking at them that may or may not be objective. This surely is a matter of learning, and carries hope for the development of a true science of man.

To a surprising extent, objectivity is something that the scientist puts on with his laboratory coat. The pressures of criticism, the traditions of methods, force a working objectivity while the scientist is dealing with the data of his specialty. But when he takes off the coat, and looks at common human problems, talks about politics or religion or foreign policy, his attitudes are apt to be just as subjective as those of anyone else.

I don't see why this should be so. I would expect that the rigorous training in methodology, is suspending judgement, in questioning dogma and authority, would carry over to attitudes outside of the laboratory. Surely it does carry over to some extent, but it is hardly noticeable in listening to the conversation of scientists in a mixed group of people. The cultural pressures are too strong for any man to dare to look at the world about him coldly and impartially, as he looks at the world under the lens of his microscope. It might be healthy, though, if some of us would try.

Actually, naturalists (and other scientists) are human enough even when they have their laboratory coats on. William Morton Wheeler stopped on occasion in the course of his critical examination of the habits of ants to look at his colleagues. He found about as many foibles and queer habits among the naturalists as among the ants.

The motivations of naturalists, as Wheeler observed, are hardly purer than those of any other group. They are not money grubbers in the ordinary sense because by taking up science in their youth, before they had learned the hard facts of their culture, they eliminated themselves from the lucrative areas of our economy, like law, medicine, business administration or advertising. Since they can't earn much money within the area of their profession, they quite sensibly erect a defensive barrier by professing scorn for the poor fellows who do slave for dough. Even naturalists must eat, though, and it is not too difficult for one university to hire professors from another by offering significantly higher rates of pay.

This is cynicism. In a very real way, financial motivations are subordinated to other things as a directing force for the naturalist. He likes honors and has developed an elaborate system of honorary fellowships, medals, and official positions in scientific societies to satisfy this craving. He also likes to see his name in print, which is one reason why the average naturalist writes so much (my house is made of very fragile glass, so please be careful). There is a social stigma in scientific circles to ordinary, or newspaper, publicity, so naturalists have learned to be shy of journalists; but they love to find references to their work in the papers of their colleagues. Scientific controversy can also be pretty bitter, with no holds barred; though this has been driven under cover by contemporary editors who refuse to publish "polemics." My wife, who has had a good opportunity to observe a wide variety of naturalists under all sorts of circumstances, says they are worse gossips than any class of women that she knows.

In short, naturalists are human beings, even the greatest of them. Darwin seemed to me a far more understandable figure after I had read his letters and found that he was pleased by praise and hurt by unkind criticism; that he was glad when Agassiz, his great opponent, made a fool of himself; that he was interested in the sales of his books.

There has been a general tendency in the biography of scientists to make them into something out of this world. Science itself could be better understood, I think, if the great scientists were studied more as men: an almost untouched field of biographical research.

THE AMATEUR NATURALIST

The usual distinction between amateur and professional—whether work is carried out for the love of it or to gain bread and butter—is not very useful in natural history. Every naturalist works at the science because he likes it. The external rewards, such as money and glory, are too feebly developed to interest people who are not already, for some other reason, strongly attracted to the subject.

Much of the basic work of natural history has been carried out by people who did not earn their living as naturalists, yet natural history was as much their profession as it is mine or that of any other scientist who has to depend on someone to write out a monthly pay check. Charles Darwin didn't earn his living as a naturalist, but the study absorbed all of his energies. He did not, in any real sense, have "amateur" status.

It would be more useful, in natural history, to distinguish between professional and amateur on a basis of whole time or part time study, and the kind of amateur that I want to discuss in this section is the person who can give only part time to natural history, for whom natural history is a hobby.

With hobby we again get involved with definition and connotations. I have come to dislike the word because somehow I associate it with killing time, a particularly venal sort of murder. The conventional methods of killing time, like solitaire, movies and detective stories, are not ordinarily listed as hobbies. But more and more, especially in America, hobbies are advertised as a therapeutic measure for papa, to keep him from dying of boredom. Whether papa dies of boredom or kills time instead is perhaps academic; but natural history need not be involved in either case.

The amateur naturalist that interests me is the fellow who might have been a professional, but got suppressed in the course of education, or sidetracked by the brutal necessity of finding bread and butter with maybe some jam. Their number is legion, and I think they would gain in satisfaction and pleasure if their part-time activities were oriented within the framework of science. It may not be possible nowadays to make revolutionary scientific discoveries on week ends home from the office, but a great deal of scientific spade work could be done on such a basis, with benefit both to science and to the amateur.

I would distinguish between the "amateur naturalist" and the "nature lover." I think the scorn that most scientists have for the nature lover comes from the disoriented, emotional attitude of such people. For the scientist it is a perversion, sterile and meaningless. And I suspect that the nature-lover complex results from the frustration of more fruitful, significant outlets, from a failure to realize the greater satisfaction that could come from trying to understand nature instead of just gushingly loving her. Maybe, of course, I am queer and not the nature lovers; maybe something is missing from my character that makes it impossible for me to look at a bird (or a picture, for that matter) with emotional satisfaction without trying to understand what I am looking at, how it works, how it got that way, where it is going.

It seems to be a general principle of the human mind, though, that pleasure and satisfaction are increased through understanding.

A Bach fugue may seem a lovely succession of sounds even to the musically ignorant, but the pleasure of listening is unquestionably increased by some knowledge of themes and counterpoint. Surely the same applies to the iris in the garden, the robin in the hedgerow or the shells along the beach.

The commonest forms of amateur natural history in the United States are probably gardening, bird watching, the maintenance of aquarium fish, and nature photography. Things like butterfly and shell collecting lag rather far behind. This is a pretty miscellaneous collection of activities, but all of them give a foundation that might be used for building up a real understanding of the scientific method; all of them might serve as a basis for real contribution to the advancement of science.

The bird watchers are the best organized of these groups, and they certainly seem to get a great deal of satisfaction out of their activities. They have also accumulated an impressive amount of factual information through their dedication to annual censuses, their charting of migration routes, and so forth. They haven't, it is true, gained much standing in the academies of science through these activities, but I suspect this is as much the fault of the scientists as it is of the bird watchers. Somehow ornithologists have got into a rather low rank in the peck-order among scientists, probably because of some behavior pattern among the professionals. But the peck-order among scientists is a separate problem, out of bounds here.

The amateur insect collectors have also made considerable contributions to science. It is possible to take a limited group of insects, such as a family of butterflies, beetles or flies, and to become a thorough master of their classification, habits and anatomy, as a part-time activity. It seems to me an infinitely more rewarding part-time activity than collecting postage stamps.

Of all of these, the people with gardens and the people with aquaria are overlooking the most golden opportunities, because they have perfect setups for using the experimental method. The gardener can investigate plant nutrition, environmental effects, competition, strain behavior, fertilization mechanisms—he has a whole biology laboratory with his seeds and his soil. The man with an aquarium (or better, with several aquaria) has an even better biological laboratory, because the whole environment is under his control.

This is turning into a missionary tract. But before stopping to get back to the main theme of the book, and to its final chapter, I want to tackle one other group of amateurs—the photographers.

Photography is a technique rather than an end in itself. This again may be classed as prejudice, but one that I share with quite respectable people. Bernard de Voto, in his column in *Harper's Magazine* once expounded at some length on the superiority of illustrative photography to art photography. It is an invaluable illustrative aid in all fields of science, from bird watching (the bottom of the peck-order) to nuclear physics (momentarily at the apex of the peck-order). I should think the amateur photographer would get an increased satisfaction out of his hobby if it were directed toward the recording or illustration of some series or class of events. The children serve this purpose while they are around and growing up, but with the children gone, nature photography would seem to be the next most easily available field.

Plants are relatively easy, because they stay put; but the photography of insects, birds, mammals and things of that sort offers endless scope for ingenuity. It is, I know, a well exploited field already; but what I am advocating here is that the photographer learn also about the organisms that he sets out to photograph, so that he can intelligently illustrate their behavior, habits and habitats. He would in no time find himself to be an entomologist with camera, rather than a photographer hunting for an angle. I think he would be happier and more productive, though of course maybe the bugs would just contribute to his misery.

At any rate, it was partly with the hope of converting some of these heathen amateurs that I started to write this book. The next and final chapter, on the methods of science, will perhaps show that these methods are nothing esoteric, that both scientific observation and experiment are based on very common human qualities which we all have shared, at least in childhood.

CHAPTER XVIII

Tactics, Strategy and the Goal

ALAN GREGG, the director for medical sciences of the Rockefeller Foundation, has made a distinction between tactics and strategy in medical research—a distinction that seems to me to have general application in science. "Strategy," he said, "is the art of deciding when and on what one will engage his strength, and tactics is the skill, economy, promptitude, and grace with which one utilizes his strength to attain the ends chosen by strategy."

Which is a nice adaptation of a military distinction. Further analogies between science and war, however, must be handled with care. In war there is an explicit goal, the defeat of the enemy; whereas in science the goal, if definable, is hardly explicit. This difference invalidates our common references to the war on ignorance or superstition, or the war against disease. It is even more misleading, I think, to refer to the mobilization of scientists for the attack on cancer; to say that the battle against yellow fever has been won; to have the physicists rallying under the banner of Einstein.

There is something wrong with all such comparisons. I can see the distinction between strategy and tactics, the selection of particular objectives and the utilization of methods adapted to the achievement of those objectives. But I balk at being mobilized, at fighting a battle, at rallying under any banner. Such comparisons are more than mere figures, they are dramatizations, misleading dramatizations that somehow twist the whole spirit of inquiry.

Science is not a war. It is rather a search. Though a queer sort of search, since we cannot be very positive about what the scientists are looking for. Once they quite clearly had the idea that they were looking for some absolute called Truth. But this vision of a Holy Grail has gradually faded, and scientists are now content with more limited objectives. They are seeking an "understanding" of natural phenomena and they have come to recognize that the precise nature of understanding will vary with the subject, with the individual, and with the stage of development of the subject and the individual.

An attempt at stating the objectives of science gets us back to the problem of definition, which formed the starting point of this book. Science, Conant said, "emerges from the other progressive activities of man to the extent that new concepts arise from experiments and observations, and that the new concepts in turn lead to further experiments and observations." This describes a method, but it leaves unstated what the method is for. The omission in Conant's definition is significant, because science is a method; and it is characteristic of scientists to be completely absorbed in the method, in the search, perhaps unconsciously making it an end in itself. If they do stop their searching to wonder why, they feel that in so doing, they have stopped being scientific to become philosophical.

I should like to close this book on such a philosophical level, but first to give a brief review of tactics and strategy, of the scientific method, as it is illustrated by natural history. The main part of this book has been concerned with the various subjects of inquiry that are included under natural history with, in the last two chapters, some account of the relation of these subjects to human economy, and of the characteristics of the people, the naturalists, who develop the science. Methods, certainly, have been mentioned in describing the various subjects, but not in any systematic manner.

The terms of Conant's definition—observations, experiments and conceptual schemes—serve conveniently as topics for a section of the tactical methods of science, though it may be useful to add a separate topic on instruments. After this review of tactics, we can give some consideration to strategy, and end with a note of speculation on the goal.

TACTICS: THE OBSERVATIONAL METHOD

Much of natural history, much of science, is descriptive, the accumulation of observations. This seems easy. But the slowness with which descriptive science has developed makes it plain that the process is not as simple as it seems.

Observation and description depend on words. The beginning of science, like the beginning of almost all distinctively human activities, thus lies in the development of language. This must have been a very slow process, whose history is forever lost in the long stretches of pleistocene time. Language has reached a high degree of development in even the most backward of contemporary cultures. In all known cultures, symbols have been developed for different kinds of sense perceptions, like colors, shapes, textures; objects are sorted into categories, like plants, rocks, birds; methods of expressing action have been developed; and of limiting description with qualifying phrases. Different contemporary languages differ greatly in the degree to which abstractions and language mechanisms are developed, but the simplest language is a tremendously complex thing compared with other animal signaling systems.

Mathematics is a special form of language, one that has been particularly useful to science because its grammar is precise and its symbols free from the encrustations of connotations that hamper our ordinary vocabulary. Almost all of the apparatus of mathematics has developed within historic times—inextricably bound up, indeed, with the development of science—so that we can see something of how the process occurred. And parts of this apparatus have passed over into our ordinary vocabulary, to form a part of the symbolic equipment that we all use in everyday thinking. The history of one of these parts may then serve as an illustration of the difficulty and slowness with which all of our methods of symbolic representation must have developed.

Our present system of numbers involves three ideas that seem simple enough, but that were tremendously difficult to come by: the use of a limited series of symbols (10); the principle of local value, whereby the order of magnitude of the symbol is indicated by its position (the 9 in 1948 designating hundreds because of its place); and the use of a symbol, zero, to indicate the absence of a quantity or unit.

All three of these ideas early form part of the thinking equipment of every child in our culture, and we cannot imagine the development of our economy or science without them. Yet, in the history of man, they are very recent discoveries.

The idea of zero, fundamental to our contemporary method of number representation, did not occur to the Egyptians, Greeks, Romans or Chinese. As far as we know, the idea of zero, the idea of treating the absence of a quantity with a particular symbol like the symbols for specific quantities, has occurred three times in the history of man. The Babylonians made the invention about 500 B.C., but it was not taken up by suosequent cultures, perhaps because it formed part of an awkward system of calculating with sixties instead of with tens. The Mayans made the discovery at about the beginning of our Christian era in connection with their calendar calculations. And the Hindus made the discovery, in about the year 500. The Hindu idea reached Europe sometime in the twelfth century, along with "Arabic" numerals, and slowly, during the following centuries, gained acceptance.

These heathen Arabic symbols met tremendous official opposition. For a long time it was illegal to use them for keeping accounts or making calculations. Man in the thirteenth and fourteenth centuries was just as pigheaded about any change, about anything different, as he is today. Looking at the record of man's conservatism, it is difficult to imagine how he ever managed to develop his languages into anything more meaningful than the ancestral monkey chatter. Some people, I suppose, would say that the increase in meaningfulness is debatable.

Of course this language business works two ways. We cannot observe and describe a thing or event until we have formulated the necessary symbols. But once we have adopted the symbols, our observations and descriptions tend to be circumscribed and directed by these very symbols.

The untangling of these relations has come to be the province of a special "science" called semantics; anyone who doubts the complexity and difficulty of the subject need only try to read one of the textbooks of this science. I am an agnostic about semantics, suspecting that the semanticists themselves have got tangled up with words. It is like trying to get off a large sheet of very sticky flypaper, or trying to unwind a tangle of Scotch tape.

The semanticists proclaim a new discovery: that words are not things. But I suspect that every scientist has long realized this; at least, insofar as he has succeeded in contributing to science, he has done so by handling his words with care, with due regard for precision of meaning and order of abstraction. Though I must admit that all of us, scientists or not, become at times the victims of word magic, of word attitudes and habits handed down with our culture. Words give wings to our imaginations; but they are also the bars that keep our minds in prison.

Science depends on the accumulation of observations, which was impossible before the invention of language and limited before the invention of writing. It also depends on the diffusion and wide availability of observations, which was limited before the invention of the printing press. The absence of a method of wide diffusion of observation, I think, made science, as we know it, impossible in the ancient world. Because the observational method, insofar as it forms part of science, must be constantly checked, and continually arranged, rearranged and codified. This is only possible if the observations can be readily recorded and widely spread.

Accuracy and codification are the two necessary qualities of scientific observation. Accuracy is always relative, which is why the constant checking process is so important. It is almost automatic in science not to accept an observation as a fact until it has been checked and repeated by several people, especially if the observation is in any way unusual. And accuracy is usually cumulative, each repeated observation filling in details and increasing exactness.

Observations are of no use in science until they have been codified and organized, the beginning of the unending chain of conceptual schemes that guide further observation. This is a process of constant adjustment. Linnaeus gathered all of the observations and descriptions that he could find of the different kinds of animals and plants, and organized these observations according to a system of classification, borrowing in the system from the ideas of many predecessors. This system has conditioned the observations of all of his successors, providing a frame into which the new observations could be fitted, the frame itself undergoing constant modification as perspectives changed with these additions.

All scientific observations must be fitted into such frames. I cannot, for instance, observe a bird, as a scientist, without the use of some such frame, code or organization. If I want to know what kind of bird it is, I make my observations within the taxonomic frame, according to the Linnaean system as at present used. If I want to observe the structure of the bird, I examine it within the frames of anatomy or morphology, according to the various organ systems. If I want to observe the habits of the bird, I watch it in the frame of ecology for actions that can be interpreted in terms of the previous knowledge of food gathering, sex behavior, territory and so forth of organisms in general, birds in general, and this kind of bird in particular.

Science has often been defined as organized knowledge, and the necessity for organization is particularly clear in this matter of observation. Providing frames for reference is an essential function of training in science. The difficulty of mastering the complex systems of organization is one of the factors that make specialization necessary in scientific work. I think a "trained observer" means an observer who knows his frames of reference, so that he can select the significant details and relate them to the observations of his colleagues and his predecessors, and hand them on in an intelligible form to his successors.

TACTICS: THE EXPERIMENTAL METHOD

Observation is a common trait of humanity, organized in a somewhat special way to form a part of science. The same is true of experiment. Some kinds of monkeys are much given to experiment, and the trait is well developed in all children. These experimental tendencies gradually become suppressed as we grow into adults by the pressures of our culture: which seems a shame, though it is probably necessary for the smooth functioning of our social structure.

To experiment, to try it, seems to me the natural impulse, inhibited and replaced in our education by subservience to authority, acceptance of dogma, from our parents or our leaders. If this idea is correct, the education is a very efficient process because the average adult human is little given to trying anything different, accepting the pattern of his life docilely or, if he rebels, doing so after some conventional fashion.

But random experiment doesn't make for science any more than random observation does. The experiments, to be scientific, must be organized, codified, directed, repeated, interpreted, fitted into conceptual schemes. Again it is a process that requires training to provide the necessary frames of reference.

The complications of scientific experiments come from design and interpretation. Both may require great skill, and the "beauty" of an experiment depends on the neatness and precision with which it is designed to clarify a particular point, and on the interpretation of the results in terms of the general area of knowledge to which the experiment is relevant.

Design and interpretation are closely related. It may be impossible to interpret the results of a poorly designed experiment, or the interpretation may be particularly difficult and complicated because the operations were not well planned. The results of the perfectly planned experiment, on the other hand, may be so clear and unequivocal as to require no interpretation.

I once heard Alan Gregg remark that many physiologists hardly understood the experimental method, since they failed to realize that the results of an experiment could never be wrong. That is a text worth pondering: the results of an experiment cannot be wrong. The experiment may be badly planned, so that it fails to illustrate the point intended; it may be badly carried out, with actual mistakes of procedure (adding arsenic to the diet formula instead of flour, for instance); or it may be badly interpreted. But the results themselves cannot be wrong. It is the essence of the experimental method that the results are the consequence of the materials and procedures.

If the results of an experiment differ from expectation, it may be that the expectation was unsound. Or it may be that some mistake was made, so that the procedure as carried out was different from the procedure as planned and recorded. Or it may be that some variable, some factor, some important element in the experiment was overlooked. The experiments that differ from expectation may, precisely, be the most interesting. The problem is to discover the reason for the divergence from expectation. This leads to more experiments, to the modification of conceptual schemes, to the progress of science.

Prediction might be treated as a special form of experiment, particularly important for the testing of concepts in the areas of science where the materials are difficult to bring under laboratory control. The prediction of an eclipse fifty years ahead, which is fulfilled with an accuracy of two seconds in 1945, is an impressive validation of the concept that led to the prediction, and the search for explanations of deviations in the accuracy of prediction leads to a modification of the concepts entirely comparable to the modifications resulting from deviation from expectancy in experiments.

Experimental biology is a relatively undeveloped science as compared with experimental chemistry and physics. We have been so busy with the enormous job of observing and describing life phenomena that we have had relatively little time to spare for experimenting with the modification of the observed processes. Yet the progress of our understanding is going to depend on the development of experimental methods of attacking problems like that of the mechanism of evolution, for instance. Our knowledge of adaptations, of natural selection, of population relationships, is based almost entirely on observation and inference from observation, and the designing of experimental techniques for the analysis of such problems is far from simple. The finding of appropriate experimental methods in these fields is, I think, the most important task confronting natural history, and its successful solution will surely result in spectacular advances in knowledge.

Progress will depend on a balance between field and laboratory work. Laboratory work is primarily experimental, the maintenance of organisms under controlled conditions, where single factors of the environment can be modified with respect to a known total situation. Field work is primarily observational, the description of the natural situation. Complete description is a physical impossibility, and selection and emphasis must depend on reasoning and on leads from laboratory experiment. The described field situation must also provide the leads for further laboratory test—another one of those interlocking chains.

There is also ample opportunity for the experimental method in the field. We have our total complex, our biotic community, a network of interlocking relationships impossible to completely describe. What happens when some one element is changed? What happens when some element of the community is removed; when a new population of some kind is added; when some environmental factor is modified? Ecology has so far been almost entirely a descriptive science, but this hardly means that it will always remain so.

TACTICS: SCIENTIFIC INSTRUMENTS

I have long maintained that a naturalist should be able to do good work with no materials beyond paper and pencil. When I have aired this theory to my colleagues, they have laughed at me, because I am notably fond of apparatus, and my laboratory is always cluttered up with a wide variety of more or less complex instruments. This seems to me irrelevant. As long as I can get the money to buy the instruments, I might as well have them. But if I couldn't get them, I don't think I would start to pine away; I hope, at least, that I would find things to do, observations or experiments to make, that didn't require instruments. All scientific work is easier with appropriate instrumental aids; some kinds of scientific work absolutely depend on particular instrumental aids. But science itself is an attitude, a method of thinking, not an accumulation of gadgets.

The instruments used in natural history fall into two broad classes: containers (such as glassware) and extensions or aids of the human sense organs. The containers, varied and complex as they may be, need concern us little. The laboratory building itself is a container, and its design may be an important element in the success of scientific work. So is a test tube or an insect cage or an aquarium; and success all of the way down the line may depend on the selection or design of appropriate containers for the material to be observed or experimented on.

It might be amusing to try to build up a classification of the instruments that serve as sense organ aids. It would at least serve to underline the general difficulties of classification, of arriving at useful and natural categories, of showing evolutionary trends, of describing by the synopsis of significant similarities and divergences. The basic division might be according to the sense organ aided: microscopes, telescopes and cameras for vision; microphones for hearing. The complicated apparatus of chemical analysis might then be considered a sort of extension of the tasting and smelling functions. Rulers and thermometers would have to be related to the sense of feeling.

It might be better to consider direct extensions of the senses, as with microscopes and telescopes, as one class of instruments. Another class, like rulers and thermometers, would be for the purpose of making measurements. Another class, like cameras and phonographs, would be for the purpose of making records. The indispensable paper and pencil would belong here, along with thermographs and recording rain gauges. This is absurd too, because it would put a glass mercury thermometer in one class of instruments, and a thermograph which records the temperature with a moving pen on a piece of paper in another class of instruments. Yet the thermograph and the camera are related in function, as are the thermometer and the ruler. Which at least serves to illustrate the thesis that logical classification is not easy.

As I said, science and scientific instruments are not synonymous. Yet the progress of science is all bound up with the progress of instrument manufacture and of manipulative technique. Leeuwenhoek with his little lenses opened the window into a whole new world that thousands of people, ever since, have been busily exploring. So did Robert Koch when he discovered that bacteria would grow in neat little colonies on a slice of potato. So did Newton and Leibnitz when, independently, they devised the calculus. All of these things, the microscope, bacterial culture, the calculus, are equally instruments. So are words. And the instruments and techniques are like words in the power that they confer on the man who can command them. But like words, they also have their limitations and their dangers; and one sometimes wonders who is master and who is slave.

TACTICS: CONCEPTUAL SCHEMES

Observation and experiment (aided or not by instruments) form part of science insofar as they lead to the formation of conceptual schemes; and the conceptual schemes in turn are scientific only if they lead to further observation and experiment. Conceptual schemes thus form one of the three basic tactical elements of science. But what are conceptual schemes in this sense?

Textbooks, at least in my college days, devoted a certain amount of space to the differentiation of hypotheses, theories and laws in science, and these, I take it, are the common varieties of conceptual schemes. They are supposed to represent a sort of sequence. The hypothesis is a tentative affair, suggested as an explanation for a group of observations, but liable to modification at any moment by the advance of science. The theory is the next stage, supported by a greater accumulation of evidence, held with considerable conviction not only by the author of the theory, but also by a majority of his colleagues. The law is the final stage, a verbal statement of some general truth. There may be numerous conflicting hypotheses to explain a given set of facts; there may be two or three alternative theories; but the law forms a definite statement unchallenged by alternatives. Thus we have the nebular *hypothesis* of the origin of the solar system; the *theory* of natural selection and the *law* of gravitation.

This distinction is undoubtedly useful, especially if we remember that it represents an arbitrary sequence in a graded series of approximations. It is misleading, I suspect, insofar as it makes the final stage, the law, a sort of definite goal, a final statement that smacks of absolute truth. The scientist has come to be very suspicious of absolutes. Once, perhaps, he was definitely searching for the "immutable laws that govern the universe," and he may still have a sneaking hope that he will discover a sensible pattern (in terms of his comprehension) for this universe. But even in physics, where law once was most apparent, the scientist is nowadays more concerned with the "uncertainty principle" or the "theory of probability" than he is with any absolutes. Those greatest of generalizations, the laws of thermodynamics, seem a little less like something graven on a tablet and handed down from the mountain when one reads the books of a contemporary physicist like Bridgman.

Scientific law, in short, is never final, never absolute. It is always subject to further test, always liable to modification. If it is to be called a law, it must hold for all of the observed phenomena, but the area of observed phenomena is constantly changing, constantly growing, and thus constantly challenging the applicability of the concept of the law. I wonder if it is not the greatest achievement of science that it has learned to deal always with uncertainties, with probabilities, with approximations. The scientist must deal with assumptions; he must act as if this or that were true, and yet be ready at any moment to backtrack at the discovery that this or that is not true, starting again on his explorations with some new scaffolding of concepts.

It seems a crazy structure, this patchwork of hypotheses, theories and laws: being torn down here while it is being built up there, shaking uncertainly at every breeze. Yet it works, because of its very flexibility. The solid brickwork of dogma may come tumbling down with any earthquake. The scaffolding of science may sway with the quake, but it can't fail because it is built as an adaptation to quakes, which are thus regarded merely as tests of its soundness. The strongest laws, like the most fragile hypotheses, are built of light material which, if it falls, will cause no irreparable damage. A blow on the head from a falling brick of dogma is easily fatal; but the damage from even a main beam of scientific law can be repaired with adhesive tape (though it may take a lot of tape).

The conceptual schemes of science are not limited to hypotheses, theories and laws. There are all sorts of other varieties. Classification is essentially a conceptual scheme, whether the classification is of organisms, habitats, phenomena, or of hypotheses themselves. Scientific classification then shares this tentative, approximate nature of scientific concepts: always subject to modification, growth, or even complete abandonment.

There must be a compromise with utility: if a given classification is changed every day, it becomes more of a nuisance than a help. Any attempt to make a classification rigid and inflexible, though, at once removes it from the area of science. It must, like all of the other kinds of concepts, be subject to change with the progressive accumulation of new observations and experiments. When to make the adjustments, and how radically to make the changes, is a matter of judgement, usually arrived at by a sort of informal consensus among scientists dealing with the classification. If a suggested change seems premature, most scientists working in the field involved will simply ignore it; if the time is ripe, the change will be widely and speedily adopted.

STRATEGY

Observations, experiments and conceptual schemes, then, are the tactical materials of science. What is its strategy?

If we go back to the remark of Alan Gregg, with which I started this chapter, strategy is the art of deciding when and on what one will engage his strength. It is, to use another figure, the focus of science.

Strategy could be examined, I think, at three levels: that of the individual scientist, that of particular scientific organizations, and that of science in general as a part of the social structure of humanity.

Every individual scientist is faced, from time to time, with the necessity of making strategic decisions: of selecting his field of work. To make this decision, he has to know something of his own abilities, limitations and preferences; something of the opportunities by way of equipment and other resources (including jobs) that are available; something of the general progress of science so as to judge what fields within the area of his competence most need cultivation.

Organizations, whether universities, research institutions, or governments, must make similar strategic decisions. At the organizational level, it is generally called "formulation of policy." The strategy must be decided upon, even when the objective is clearly stated. The objective, for instance, may be medical research. Is this best achieved under prevailing conditions at the time of the decision by concentration on psychological aspects of disease, or on research in human physiology, or on studies of therapeutic agents; or is medical progress handicapped by lag in general biological theory so that the long-range objective can best be served by some apparently divergent immediate policy? No organization has resources enough to be able to ignore such policy questions, and the success of the organization depends in very large measure on the wisdom of its strategic decisions.

Society, in a more or less blind way, similarly controls the strategy of science in general. It does this by the apportionment of recognition. If science in general is highly esteemed, it is apt to flourish; and the particular aspects of science that most appeal to the society are apt to flourish most. This of course again is a sort of a chain reaction, since the relative esteem of different aspects of science surely is determined in part by the personalities and productiveness of the different scientists themselves. And science, for flourishing, requires something more in the environment than esteem. We in the United States have long professed to hold science in high esteem, yet it is only during very recent years that we have begun to make contributions proportional to our wealth and population. Some necessary factor was lacking in our cultural environment.

In the years since World War II, national policy with regard to fostering science has become a matter of general public interest and of much discussion. Strategic decisions at this level may thus become more explicit and conscious. The consequences of this are bound to be far-reaching—but whether for "good" or for "evil" I don't know, and any attempt at analysis here would take us far out of bounds.

THE GOAL

I have, then, tried to describe an area of science, natural history, as an example of the whole, as a sample of the content and method. Purpose—the question of objectives—has inevitably arisen from time to time, and I have tried to dodge by pointing out that this concept belongs to the field of philosophy.

But at the same time I have deplored all fences and barriers between fields of knowledge, and one of my strongest convictions is that methods must be found of reconciling, of joining, of connecting science and philosophy. My attempt to avoid purpose, then, is really a present convenience, an effort to keep this book within the range of my ability and of the reader's patience.

The difference between science and philosophy, as I see it, is a matter of method. The one seeks for conceptual schemes through the methods of controlled observation and experimentation, while the other arrives at its concepts through introspection, through meditation, through thought processes that superficially seem very different from those of science. The final schemes, though, may be remarkably similar; and the goal, of understanding, of developing a correspondence between the external world of reality and the internal world of the mind, is surely identical.

Science, then, is a search, but without an explicit goal; a

by-product of the restlessness that seems to be innate in man. It is, by definition, a certain way of searching, but I do not see how this limits the area in which the way is to be tested.

We call the object of our searching "understanding." But understanding is not a clear, concrete, definite thing. We cannot imagine that we shall find it complete, and carefully wrapped in colored paper, at some definite date in the future like September 16th, in the year 2045. It is something we acquire bit by bit. Each bit makes us want more, but each bit also brings its satisfactions. Perhaps these little satisfactions should be sufficient end in themselves, but man-like, we always hope for more, always look to some farther horizon.

My hope is that in the process of this searching, in the course of finding bits of understanding, we may also, on the way, acquire common sense, and learn to face the world more boldly as we find it. Learn thus to live together more pleasantly. Find, perhaps, some understanding of ourselves.

APPENDIX

The Literature of Natural History

IF THIS book had been written as a scientific treatise, it would have been documented: that is, each statement would be supported by a reference indicating where the information came from. This is a tradition of all scholarship, which science shares with history, criticism and theology. It can be an awful nuisance both for the writer and the reader, but it can also be both useful and interesting. Much depends on the mechanics of the documentation.

In science, particularly, it is a basic tenet that a mere assertion carries no weight, means nothing beyond the expression of an opinion by the writer. If I write that "cats eat mice" as a bald statement, no scientist can give the statement any value beyond noting that "Bates makes the unsupported assertion that cats eat mice." If it is an original observation, I must say where, and under what circumstances, I observed cats eating mice, so that my evidence can be given proper perspective by anyone making a general survey of the diet of cats, or the predators of mice. If my assertion is second-hand, I must give my source, so that the reading scientist can check back to find the circumstances and the evidence of the man who did make the observation.

If, for some reason, I become interested in the phenomena of

285

cats eating mice, and make some observations which I intend to publish as a contribution to science, it is my duty to make a survey of what other people have written about cats eating mice, so that my observations can be compared with those of my predecessors. If many people have written about the mysophagy of felines (that's good jargon for you) I don't need to cite every previous author, but I can cite a few authors who, in their turn, have given references and summaries of the work that went before theirs. On such rests the cumulative nature of the factual raw material of science. Learning how to track down the literature, how to find out what has been done before, how to use the indexing systems and how to carry personal indexes on subjects of particular personal interest, is a very important part of any scientific education. Learning how to look it up ought to be the most important part of any education.

There are many ways of giving documentation. Perhaps footnotes are the commonest form, and I have heard "scholar" defined as "a person who, every time he opens his mouth, inserts a footnote." I think footnotes are an unmitigated nuisance. The theory is, that the casual reader will keep right on, paying no attention to the details of documentation. But with me, at least, the habit of reading footnotes is deeply ingrained, and every time I come across the superscript,[1] my eye automatically drops to the bottom of the page, and my concentration on the author's thought, always a delicate balance, is shattered.

The notes can all be put in the back of the book. But if they are interesting, this is even more awkward, and if they are not interesting, they serve little purpose. Notes in the back of the book are meant as a defense for the author against critics; but, like all such defensive measures, they are generally ineffectual.

The best system, I think, is somehow to work all of the quali-

[1] Frothingham (J. roy. Soc. Mamm., xiv, p. 412) points out that Aristotle (Hist. An., Bk. II) saw a cat eat a mouse, and that the Egyptians probably knew of such occurrences, though they worshipped cats for different reasons. If you read footnotes assiduously enough, you sometimes come across pornographic bits put in Latin.

fications and hedgings into the text, and to make the documentation a straight list of references, devising the simplest and most unobtrusive system possible for getting cross connections between text and reference for the scholar who wants to dig further into the subject. The use of such a system here was made difficult by the fact that this book was mostly written in the small Colombian town of Villavicencio with only my own library for reference; and completed with that library packed up and in storage.

The basic literature of biology is, in any event, the accumulation of serial journals; and references to these journal articles would be out of place in a book of this sort. The journals are mostly available only in technical libraries, and the articles are usually written with a vocabulary that is unintelligible except to the specialist in the narrow field covered by the journal. Anyone wanting to become familiar with any one of the special biological sciences must learn to use these journals, but that is easy enough once the particular field—bacteriology, entomology, ecology, ornithology—has been selected. The general books on these various sciences almost always contain detailed bibliographies which serve as an introduction to the periodical literature (as well as to other books) in the field.

There are many hundreds of contemporary books in English on the natural history sciences, and plainly the bibliography of the present book should serve as a guide to the reader who wants to follow up some particular aspect of these sciences in more detail. Some such guide is necessary because neither the bookseller nor the librarian can be expected to judge the relative merits of the numerous possible introductions to the particular subjects.

But the compilation of an adequate guide would be far from easy, and would inevitably reflect the particular prejudices and interests of the compiler. All I can pretend to do here is to offer a few suggestions.

Natural history books could be grouped into three classes: those designed for reading, those designed for reference, and textbooks. The books for reading run a gamut according to the background that the author presupposes in his readers, from books like the present one, that presuppose no special training, to books written by specialists for other specialists in some narrow field. Some of the most stimulating and significant biological books belong at this specialized end of the ganut: they are closed to the reader who has not mastered the vocabulary and factual accumulation of the particular specialty. On the other hand, the books that presuppose no special training, often fail to provide any ideas. They may be easily and painlessly read, but they resemble soda pop in that the flavoring is artificial, the nutritive value nil.

Books for reference run a similar gamut, but there is a more sensible correlation between assumed reader knowledge and information made available. The basic references are the encyclopedias. I have consulted the *Encyclopaedia Britannica* constantly in the preparation of this book—and I acknowledge here my debt to the authors of the various biological articles. Scholars usually consider that the *Britannica* has deteriorated from the 11th to the 14th editions, but the 14th edition is still an authoritative reference for many aspects of natural history. The 15th edition, judging by parts I have seen written by friends, will be even more useful and authoritative.

Every naturalist depends on a series of reference books for the identification of the organisms that particularly interest him, and for basic information about their habits, distribution and structure. Thus anyone interested in wild flowers in the eastern United States would always have a copy of Gray's *New Manual of Botany* at hand for reference; in butterflies, a copy of Holland's *Butterfly Book*. For conspicuous animal and plant groups, in well-known regions like Europe and North America, many such reference books are available. For less conspicuous groups, like most invertebrate families, recourse must be had to technical monographs. The technicality of the vocabulary is relatively unimportant in these books, however, because they are meant for reference, not reading, and because anyone acquiring a special interest in some particular group of organisms soon masters the vocabulary applying to their group. Such reference books, in any case, almost always include glossaries in which the technical terms are explained.

These reference books are too numerous for me to attempt to list them here. There is an extensive, annotated list of books applying to the North American biota in Comstock's *Handbook of Nature Study* which should be consulted by anyone looking for material on particular subjects.

Textbooks form a class of publications all by themselves. I gathered a shelf of biological texts, thinking to cull from them a guide for the reader of this book. But the problem defeats me. Each text seems duller and more arid than the last. I have asked a variety of professor friends for suggestions, but the text one likes the other calls "lousy." I have protested at the uniform dullness of these texts, and asked why more interesting ones weren't written. The answer of one professor is perhaps illuminating. He said that he liked to stimulate and interest his students through his lectures, and that the text should serve for factual reference only. I can't judge this, since I have never taught, but I should think it would be nice to have both an interesting text and an interesting teacher.

In the end, I have decided to list no texts except the ones that I have used extensively for reference while writing this book. The choice of these was largely a matter of accident. Any textbook put out by one of the major text publishers, and written by a professor in a major university, is apt to be reasonably accurate within textbook limits, though almost sure to be dull fireside reading.

Then there is the periodical literature. This divides sharply into "popular" and "technical." I can't work up much enthusiasm for any of our "popular" natural history journals, since they all tend to stress the spectacular, reflecting the "oh my! ain't nature grand" attitude. They often purvey facts, rarely ideas. The *Scientific*

Monthly, an official publication of the American Association for the Advancement of Science, is supposed to interpret science for the layman, but has succeeded mostly in helping to interpret science to scientists—an important enough job in itself. To be enthusiastic about journals interpreting science in nontechnical language, I have to turn to England, to things like *Endeavour* and *Discovery,* which aren't available on American magazine stands. I think it is a safe generalization to say that the English scientist is, on the average, a much better writer than his American colleague, which may reflect some difference in our educational systems.

There are two general technical journals that I think should be mentioned: *Biological Reviews,* published at Cambridge, England; and the *Quarterly Review of Biology,* published at Johns Hopkins in Baltimore. The articles in both are apt to be highly technical, but *Biological Reviews,* especially, has published a series of extraordinarily able articles over the years summarizing advance in various specific aspects of biological knowledge. The *Quarterly Review of Biology* publishes articles that seem to me of less general interest; but it publishes splendid book reviews that pull no punches and that serve as the best guide through the forest of biological texts, reference books, and fireside reading.

In the reference list of this book, I have included citations of periodical articles only in a few cases where they seemed necessary for the documentation of my source of information.

From these various considerations, then, I end with something that is far from a balanced guide to the literature of the natural history sciences. It contains some titles that I think the reader will enjoy, as I have; a partial documentation of my sources; and a few books whose bibliographies, in their turn, should serve as useful charts for beginning exploration in the vast hinterland of biological literature.

References

Allee, W. C. *The Social Life of Animals*. New York: W. W. Norton
& Co., 1938. [Cited in Chapters XI and XII. This is the most
general of Allee's writings, full of stimulating ideas and easy
to read.]

Armstrong, E. A. *Bird Display and Behavior. An Introduction to
the Study of Bird Psychology*. New York: Oxford University
Press, 1947. [Particularly valuable because of the detailed
bibliography.]

Bates, Marston. *The Natural History of Mosquitoes*. New York:
The Macmillan Co., 1949. [This book was written for specialists, but it may serve as an example of my idea of how the
natural history sciences can be focussed on a particular group
of organisms.]

Bates, Henry Walter. *The Naturalist on the Amazon*. London: John
Murray, 1864. [Cited in Chapter XIV. A one-volume edition
is reprinted in "Everyman's Library." Bates on the Amazon
and Belt in Nicaragua are classics of natural history, peren-
nially interesting to read.]

Beebe, William. *The Book of Naturalists. An Anthology of the
Best Natural History*. New York: Alfred A. Knopf, 1944.
[Beebe's own writings also serve as an easy approach to natural
history; though scientists generally consider them too highly
colored with the nature loving, or "oh my!" attitude.]

de Beer, G. R. *Embryos and Ancestors*. New York: Oxford University Press, 1940. [Cited in Chapter VI. A stimulating book
for the biologist, but the de Beer vocabulary will stump the
nonprofessional.]

Belt, Thomas. *The Naturalist in Nicaragua*. London: John Murray,

1874. [Cited in Chapter IX. Still one of the most readable of the books by traveling naturalists; reprinted in "Everyman's Library."]

Brues, C. T. *Insect Dietary. An Account of the Food Habits of Insects.* Cambridge: Harvard University Press, 1946. [Cited in Chapter III. Contains much information about things other than diet, and a detailed bibliography.]

Chase, Stuart. *The Tyranny of Words.* New York: Harcourt, Brace & Co., 1938. [This seems to me the easiest introduction to semantics: a subject I have libeled several times in the present book.]

———— *The Proper Study of Mankind. . . . An Inquiry into the Science of Human Relations.* New York: Harper & Bros., 1948. [Cited in Chapter XII; incidentally, a good introduction for the layman to some aspects of the social sciences.]

Comstock, A. B. *Handbook of Nature-Study.* Ithaca: Comstock Pub. Co., various editions, 1911 to date. [This has proved a very successful introduction to "Nature Study" for the eastern United States. It is arranged as a series of lessons on a range of subjects from birds to stars. It contains an extensive, annotated list of books on natural history subjects.]

Conant, James Bryant. *On Understanding Science.* New Haven: Yale University Press, 1947. [Cited in Chapters I and XVIII. This book stimulated me to write the present one, and it is, I think, a book that every educated man should know. Friends to whom I have recommended it sometimes report it difficult to read, but it is worth the effort.]

Cutright, P. R. *The Great Naturalists Explore South America.* New York: Macmillan, 1940. [A good compilation on the conspicuous animals of South America.]

Darlington, P. J., Jr. "The origin of the fauna of the Greater Antilles, with discussion of dispersal of animals over water and through air." *Quarterly Review of Biology,* vol. 13, 1938, pp. 274–300. [Cited in Chapter XIII.]

———— "The geographical distribution of cold-blooded vertebrates." *Quarterly Review of Biology,* vol. 23, 1948, pp. 1–26; 105–123. [Cited in Chapter XIII.]

Darwin, Charles. *On the Origin of Species by means of Natural*

Selection, or the Preservation of Favoured Races in the Struggle for Life. London: John Murray, 1859. [Reprints of all of Darwin's books are easily available, especially of the "Origin," the "Descent of Man" and the "Voyage of the Beagle." The "Beagle" forms the easiest introduction to Darwin's writings, since most people find the "Origin" slow going. The best biographical introduction to Darwin, I think, is that of Geoffrey West cited below.]

Daubenmire, R. F. *Plants and Environment. A Textbook of Plant Autecology.* New York: J. Wiley, 1947. [Consulted frequently; it is, perhaps, easier reading than most plant ecology texts.]

Dobzhansky, Theodosius. *Genetics and the Origin of Species.* New York; Columbia University Press, 2nd ed., 1941. [Cited in Chapter XIV. This is one of the most important post-Darwin books on evolution; it is not easy reading because of the specialized vocabulary of genetics.]

Dunbar, C. O. *Historical Geology.* New York: Wiley, 1949. [The predecessor of this (by Schuchert and Dunbar) served as my most frequent reference for paleontology.]

Elton, Charles. *Animal Ecology.* New York: Macmillan Co., 1936. [Cited several times. To my mind, the best introduction to animal ecology; short and easy to read as well.]

———— *Voles, Mice and Lemmings. Problems in Population Dynamics.* Oxford: The Clarendon Press, 1942. [Cited in Chapter XII.]

Errington, P. L. "Predation and vertebrate populations." *Quarterly Review of Biology,* vol. 21, 1946, pp. 144–177; 221–245. [Cited in Chapter XI.]

Fisher, R. A. *The Genetical Theory of Natural Selection.* Oxford: The Clarendon Press, 1930. [Cited in Chapter XIV. Tough going.]

———— *The Design of Experiments.* Edinburgh: Oliver & Boyd, 4th ed., 1947. [A standard reference on the subject, not arranged for fireside reading.]

Frobisher, Martin. *Fundamentals of Bacteriology.* Philadelphia: W. B. Saunders Co., 3rd ed., 1944. [This is the bacteriology text I consulted most; it is one of many good ones.]

Garretson, M. S. *The American Bison. The Story of its Extermina-*

tion as a Wild Species and its Restoration under Federal Protection. New York: N. Y. Zoological Society, 1938. [Cited in Chapter XII.]

Goldschmidt, Richard. *The Material Basis of Evolution.* New Haven: Yale University Press, 1940. [Cited in Chapter XV; definitely tough reading.]

Griscom, Ludlow. *Modern Bird Study.* Cambridge: Harvard University Press, 1945. [Cited in Chapter XII. There are many books on birds, and I don't know them well enough to venture recommendations. This one is easy to read, and written by an expert ornithologist.]

Hamilton, W. J., Jr. *American Mammals. Their Lives, Habits, and Economic Relations.* New York: McGraw-Hill, 1939. [A good introduction to the natural history of mammals.]

Hegner, R. W. *Parade of the Animal Kingdom.* New York: Macmillan Co., 1935. [A good introduction to the diversity of animal life.]

Henderson, L. J. *The Fitness of the Environment. An Inquiry into the Biological Significance of the Properties of Matter.* New York: Macmillan Co., 1913. [A "must" for all interested in the philosophy of biology.]

Hesse, R., W. C. Allee and K. P. Schmidt. *Ecological Animal Geography.* New York: John Wiley & Sons, 1937. [Frequently consulted in preparing the present book. A completely revised edition is in preparation: basic reading for all naturalists.]

Hogben, Lancelot. *Mathematics for the Million.* New York: W. W. Norton & Co., 1937. [Cited in Chapter XVI. This contains many stimulating observations on the history of science in general as well as on mathematics. Hogben is primarily a biologist, belonging to the "H" group whose names guarantee smooth reading: Hogben, J. B. S. Haldane and Julian Huxley.]

Hooton, E. A. *Man's Poor Relations.* New York: Doubleday, Doran & Co., 1942. [An easy, but very completely documented introduction to all aspects of primate biology.]

Howard, H. E. *Territory in Bird Life.* London: John Murray, 1920. [Cited in Chapter XI; historically, an important book, and one that is also stimulating reading.]

Huxley, Julian. *Evolution. The Modern Synthesis.* New York: Harper & Bros, 1942. [Cited in Chapter XIV. This is probably the best single book for the serious reader who wants a survey of contemporary biological thought on problems of evolution, but it is not written with Huxley's usual fluency.]

Kendeigh, S. C. "Measurement of bird populations." *Ecological Monographs,* vol. 14, pp. 67–106, 1944. [Cited in Chapter XII.]

Kluckhohn, Clyde. *Mirror for Man. The Relation of Anthropology to Modern Life.* New York: Whittlesey House, 1949. [Cited in Chapter XVII. This book is a sort of introduction to the "natural history of man."]

Loeb, Jacques. *Forced Movements, Tropisms and Animal Conduct.* Philadelphia: J. B. Lippincott Co., 1918. [Cited in Chapter XI.]

MacGinitie, G. E. and Nettie MacGinitie. *Natural History of Marine Animals.* New York: McGraw-Hill Book Co., 1949. [This book came to hand too late to be used in the preparation of the present manuscript, except for a quote in Chapter XIV, but it looks like a good introduction to marine biology, written in textbook style.]

Matthew, W. D. *Climate and Evolution.* New York: 2nd ed., special publication of the N. Y. Academy of Sciences, 1939. [Cited in Chapter XIII. A book for the specialist, but one that has had much influence on contemporary biological thought.]

Mayr, Ernst. *Systematics and the Origin of Species.* New York: Columbia University Press, 1942. [Cited in Chapter XV. Deals chiefly with the mechanism of speciation, stressing geographical factors. Though written for biologists, it is not too difficult for lay readers.]

Monge, Carlos. *Acclimatization in the Andes.* Baltimore: Johns Hopkins Press, 1948. [Cited in Chapter XIV.]

Nice, Margaret. "The role of territory in bird life." *American Midland Naturalist,* vol. 26, 1941, pp. 441–487. [Cited in Chapter XI.]

Nordenskiold, Erik. *The History of Biology.* New York: Alfred A. Knopf, 1928; re-issued by Tudor Pub. Co., 1935. [Agreed to be excellent, and frequently consulted during the preparation of this book; it is not light reading.]

Osborn, Fairfield. *Our Plundered Planet*. Boston: Little, Brown, 1948. [Cited in Chapter XVI. Along with Vogt's book, this documents the contemporary dim view of resources exploitation.]

Osborn, H. F. *From the Greeks to Darwin*. New York: Charles Scribner's Sons, 2nd ed., 1929. [Cited in Chapter XV.]

Pearl, Raymond. *The Natural History of Population*. New York: Oxford University Press, 1939. [Cited in Chapter XII.]

Read, Carveth. *The Origin of Man and His Superstitions*. Cambridge: The University Press, 1920. [Cited in Chapter XI. A book that is not taken seriously by anthropologists.]

Rogers, J. S., T. H. Hubbell and C. F. Byers. *Man and the Biological World*. New York: McGraw-Hill, 1942. [This is the biological textbook I have consulted most frequently, perhaps because it was written by friends; the second part of it, on paleontology and ecology, is particularly good.]

Romer, A. S. *Man and the Vertebrates*. Chicago: University of Chicago Press, 3rd ed., 1941. [An interesting textbook on vertebrate biology as seen by a paleontologist. In preparing the present book, I have more frequently consulted a stiffer text by the same author, "Vertebrate Paleontology" (Univ. of Chicago Press, 2nd ed., 1945).]

Russell, Bertrand. *The Scientific Outlook*. New York: W. W. Norton & Co., 1931. [A "must" for anyone interested in the philosophy of science. Like much of Russell, it is so easy to read that one scarcely realizes the profoundness of the thought.]

Sarton, George. *The Life of Science. Essays in the History of Civilization*. New York: Henry Schuman, 1948. [The idea that the history of science serves as a gateway to its understanding is rapidly gaining ground; these smoothly written essays serve as a nice introduction to that history.]

Simpson, G. G. *Tempo and Mode in Evolution*. New York: Columbia University Press, 1944. [Cited in Chapter XV. The book is technical, written for biologists, but Simpson has an easy style that smooths the way.]

Toynbee, A. J. *A Study of History*. (Abridgement by D. C. Somervell) New York: Oxford University Press, 1947. [Cited in Chapter VIII; scientists in general are apt to disagree with Toynbee's interpretations of history on several important points.]

Vogt, William. *Road to Survival.* New York: William Sloane Associates, 1948. [In reading this book, I found myself constantly provoked to disagreement over minor points. But it is meant to be a provoking book, and if the subject (conservation of resources) had been approached calmly, with perspective and balance, the author might have failed to reach the wide audience that we all agree should be aroused to the urgency of the conservation problem.]

Wallace, A. R. *The Geographical Distribution of Animals.* London: Macmillan & Co., 2 vols., 1876. [Cited in Chapter XIII.]

Weaver, Warren. "Science and complexity." *American Scientist,* vol. 36, 1948, pp. 536–544. [Cited in Chapter XII.]

Wells, H. G., Julian Huxley and G. P. Wells. *The Science of Life.* New York: Doubleday, Doran & Co., 1931, 2 vols. [This is an excellent summary of the biological sciences written for the layman.]

West, Geoffrey. *Charles Darwin. A Portrait.* New Haven: Yale University Press, 1938. [An excellent biography.]

Wheeler, W. M. *Ants: Their Structure, Development and Behavior.* New York: Columbia University Press, 1910. [Cited in Chapter IX. This is still the standard book on the natural history of ants.]

——— *Foibles of Insects and Men.* New York, Alfred A. Knopf, 1928. [The chapters on the "Organization of Research" and the "Dry Rot of our Academic Biology" should be read by everyone interested in the philosophy of science. Another book by this author, *Social Life Among the Insects,* should also be on every natural history shelf.]

Willis, J. C. *Age and Area. A Study in Geographical Distribution and Origin of Species.* Cambridge University Press, 1922. [Cited in Chapter XIII.]

Yerkes, R. M. *Chimpanzees: A Laboratory Colony.* New Haven: Yale University Press, 1943. [Cited in Chapter XI.]

The Recent Literature
of Natural History

I HAVE tried to limit the following references to works in the spirit of those in Bates's excellent list, which contains many books that I still recommend highly in my field biology course at Princeton. I have concentrated on textbooks, general references, and specific ideas, rather than on natural histories by modern writers.

I dearly wish that I could add many contemporary natural historians to Bates's list. But unfortunately, most of today's acclaimed "nature writers" are what Bates refers to as "nature lovers" (pp. 253 and 264). They tend to describe just what they have seen, and perhaps to rhapsodize about it. What they see is indeed beautiful, but it is trivial in the sense that any sensitive person in the same environment would probably see the same things. Very rare are those who observe nature perceptively, and think about the ecological patterns and evolutionary relationships among the creatures and vegetables that they see. Many of the best modern observers and thinkers are specialists, as Bates was a mosquito specialist, but few of them have the broad perspective he had.

It is also hard to find textbooks and scientific periodicals that are engaging for anyone but a specialist. To Bates's list of journals, I would add *Natural History, American Scientist, New Scientist,* and *Trends in Ecology and Evolution.* The last of these is particularly valuable for giving signposts to the technical literature in articles that are intelligible to the general public.

It is much easier to recommend field guides, which have prolif-

299

erated since the 1950s. *A Field Guide to . . . (The Peterson Field Guide Series)*, from Houghton Mifflin of Boston, Mass., is perhaps the best all around for identification. The books are a handy size, and have technical keys for identification, descriptions including similar species, and pictures of varying realism, each with a little arrow pointing to a diagnostic feature. (Roger T. Peterson copyrighted the last feature, preventing all other field guides from using it.) Natural history is confined to introductory sections in the Peterson guides, but some of those sections are excellent. Golden Press of New York has two series, *A Guide to Field Identification* . . . , which has excellent paintings to leaf through once the major groups of species have become clear, and *Golden Guide* . . . , which is the best series for neophytes. The tiny *Golden Guides* have a superb selection of common species, along with charming and accurate paintings, with brief notes on natural history. From Collins of London comes a new kind of field guide, the *New Generation Guide to . . . of Europe*, with an emphasis on natural history. The closest analog for North America, Kricher and Morisson's *Field Guide to Eastern Forests*, is listed below. There are also several large books that have excellent snippets of natural history for a wide range of organisms, in particular the books edited by Chinery and by Wernert listed below.

—Henry S. Horn

Angel, Heather, et al. *The Natural History of Britain and Ireland*. London: Mermaid, 1985 reprint of 1981 edition. [Angel's color photographs are superb, as is the text.]

Angel, Heather, and P. Wolseley. *The Water Naturalist*. New York: Facts on File, 1982. [This is an excellent guide to observing aquatic populations and ecosystems. It complements Durrell and Durrell, which is listed below.]

Bartram, William. *Travels through North & South Carolina, Georgia, East & West Florida, the Cherokee Country, the Extensive Ter-*

ritories of . . . etc. New York: Dover, 1955 reprint of 1791 edition. [This is at once a classic of exploration, of natural history, and of expeditionary purple prose.]

Bates, Marston. *Gluttons and Libertines: Human Problems of being Natural.* New York: Random House, 1967. [Much of this is outdated, and some is now wrong, but because of Bates's unusual combination of perception and kind humor, it is still a useful antidote to much of what is being written today about biological perspectives on human behavior.]

Beebe, Charles William, ed. *The Book of Naturalists: an Anthology of the Best Natural History.* Princeton: Princeton University Press, 1988 reprint of 1944 edition. [See Bates's notes.]

Belt, Thomas. *The Naturalist in Nicaragua.* Chicago: University of Chicago Press, 1985 reprint of 1874 edition. [See Bates's notes.]

Bonner, John T. *Evolution of Culture in Animals.* Princeton: Princeton University Press, 1989 reprint of 1980 edition. [This is a popularization of technical work that has shown clear analogs of crucial aspects of human cultural evolution in other animal populations. John Bonner has also written about aspects of developmental biology that have profound implications for natural history. He has considerable insight into the evolution of complexity, which Bates refers to briefly on p. 223. It is particularly interesting to compare Bonner's ideas about the evolution of complexity in two books, one written recently and the other at the time of Bates's book, when the full implications of the revolution in molecular biology were just beginning to be explored by natural historians. Bonner's appropriate books are *The Ideas of Biology*, New York: Harper & Brothers, 1962, and *The Evolution of Complexity by Means of Natural Selection*, Princeton: Princeton University Press, 1988.]

Brooks, Paul. *Speaking for Nature: How Literary Naturalists from Henry Thoreau to Rachel Carson Have Shaped America.* San

Francisco: Sierra Club, 1983 reprint of 1980 edition. [This is a good sample of natural history writing with an excellent balance between interpretive text and quotations.]

Chinery, Michael, gen. ed. *The Natural History of Britain and Europe*. London: Kingfisher, 1982. [Here, in a fat but handy volume, are attractive paintings of most common organisms, along with brief notes on their habits and habitats.]

Colinvaux, Paul. *Why Big Fierce Animals Are Rare*. Princeton: Princeton University Press, 1978. [This introduction to some important conceptual issues in ecology is especially appropriate if it is your first exposure to the field.]

Comstock, Anna Botsford. *Handbook of Nature Study*. Ithaca: Comstock, Cornell University Press, 1986 reprint of 1911 edition. [Bates has already listed and described this book, but I can't help noting its indirect effect on my education. My mother used Comstock's *Handbook* as an undergraduate at Cornell. Some of Comstock's ideas and style are dated (as were some of my mother's), but the observations and enthusiasm are timeless. My mother wrote her master's thesis in Potato Science, and it has been checked out of the library once, by the government of Morocco . . . but I digress.]

Darlington, P. J. *Biogeography of the Southern End of the World*. Cambridge, Mass.: Harvard University Press, 1965. [This updates Bates's reference to Darlington. It is particularly instructive because it was written just as the first incontrovertible geological evidence of continental drift was accumulating. Darlington's discussion of continental drift shows extraordinarily fine perception and judgment for its time.]

Darwin, Charles. [Darwin was one of the best naturalists of all time. You could profitably read everything that he wrote. Particularly good facsimiles of first editions are: *Origin of Species*, introduced by E. Mayr, Cambridge, Mass.: Harvard University Press, 1964; and *The Descent of Man and Selection in Relation to Sex*, introduced by J. T. Bonner and R. M. May, Princeton: Princeton University Press, 1981. The University of Chicago Press has reprinted several of Darwin's lesser known works including *The Collected Papers of Charles Darwin*, Paul Barret, ed., foreword by Theodosius Dobzhansky (1977); *The Expression of the Emotions in Man and the Animals*, preface by Konrad Lorenz (1965); *The Formation of Vegetable Mould, through the Action of Worms, with Observations on Their Habits*, foreword by Stephen Jay Gould (1985); *Different Forms of Flowers on Plants of the Same Species*, foreword by Herbert G. Baker (1977); and *The Various Contrivances by Which Orchids are Fertilised by Insects*, foreword by Michael T. Giselin.]

Dawkins, Richard. *The Selfish Gene*. Oxford: Oxford University Press, 1976. [This is an eloquent popularization of the notion that advantages in replication of genes can take precedence over hypothetically advantageous characteristics of individuals, populations, species, and evolutionary lineages, because the only characteristics that can evolve by natural selection are those that are genetically inherited. This concept undermines simple statements about evolution favoring survival of populations and species, and calls for greater care in statements like those that Bates makes on pp. 108, 154, 217, 223, and 236. *The Selfish Gene* has been criticized for slightly overstating the primacy of genes over the individuals that carry them, but Dawkins himself has been active in such constructive criticism. So I expect that his second edition, due in 1989, will become the definitive statement.]

Durrell, G., and L. Durrell. *The Amateur Naturalist*. New York: Knopf, 1983, paperback ed. 1989. [This is the best of a good lot of recent guides to field techniques. It is superbly illustrated. It is written in English, but translation to North American should be easy.]

Ehrlich, Paul R. "Population Biology, Conservation Biology, and the Future of Humanity." *BioScience*, vol. 37, no. 10, 1987, pp.

757–763. [Read this. Tell your friends. Send it to your representatives in government.]

Ehrlich, Paul R., Anne H. Ehrlich and John P. Holdren. *Ecoscience:
Population, Resources, Environment.* San Francisco: W. H.
Freeman, 1977. [This excellent textbook is about due for revi-
sion. Much of the new information will doubtlessly be reviewed
in Ehrlich and Ehrlich's *The Population Explosion*, due in 1990
from Simon & Schuster.]

Ehrlich, Anne H., and Paul R. Ehrlich. *Earth.* New York: Franklin
Watts, 1987. [This book paints a shocking and depressing picture
of past and present global human ecology, and sets a hopeful
agenda for the very near future.]

Farb, Peter. *The Face of North America.* New York: Harper & Row,
1963. [Farb writes a fine introduction to the geology and ecology
of North America. Some of the details of his geology need re-
evaluation in the light of new notions about continental drift, and
his discussion of plant succession portrays more regularity than
present ecologists think they find. His list of places to visit is superb.]

Forsyth, Adrian, and Ken Miyata. *Tropical Nature: Life and Death in
the Rain Forests of Central and South America.* New York: Scrib-
ners, 1984. [This is one of the few recent natural histories I have
read that recaptures both the perception and the enthusiasm of the
explorers of the last century.]

Friday, Adrian, and David S. Ingram, eds. *The Cambridge Encyclopedia of Life Sciences.* Cambridge, England: The University
Press, 1985. [This is the most readable substitute for a general
biology textbook that I have found. It is very weak on molecular
biology.]

Gould, James L., and Carol Gould. *The Honey Bee.* New York: W. H.
Freeman, *Scientific American Library*, 1988. [On pp. 40 and 159
Bates vastly underestimates the mental powers of invertebrates.
Gould and Gould review recent evidence of the extraordinary
abilities of bees to learn and also to recognize particular pieces of

real estate. I have observed behavior of similar complexity among butterflies, and I suspect that comparable intelligence will be found to be widespread among animals. Incidentally, the *Scientific American Library* has many other books on special topics of interest to natural historians. All are exceedingly well written and copiously illustrated.]

Gould, Stephen Jay. *Wonderful Life: The Burgess Shale and the Nature of History*. New York: Norton, 1989. [This book adds astounding new evidence to Bates's story of the major phyla of animals in Chapter III. Gould writes a regular column in *Natural History* magazine that is always a marvelous combination of arcana, insight, and rhetoric. Many of his essays address the difficult issues of "purpose" and "progress" in evolution, issues to which Bates refers elliptically on pp. 57, 210, and 250–251. Several excellent collections of Gould's *Natural History* essays have been published by Norton: *Ever Since Darwin*, 1977; *The Panda's Thumb*, 1980; *Hen's Teeth and Horse's Toes*, 1983; and *The Flamingo's Smile*, 1985.]

Harris, Larry D. *The Fragmented Forest: Island Biogeographic Theory and the Preservation of Biotic Diversity*. Chicago: University of Chicago Press, 1984. [This is a lucid development of practical applications that owe equal debts to old fashioned natural history and to the latest in theoretical ecology.]

Jolly, Alison. *The Evolution of Primate Behavior*. New York: Macmillan, 2nd ed., 1985. [This is an eloquent and penetrating introduction to the natural history of us and our closest relatives. The beginning of the first chapter is a compelling argument for biological conservation.]

Keeton, William T., James L. Gould, and Carol G. Gould. *Biological Science*. New York: Norton, 4th ed., 1986. [I must recommend this particular general biology text, not only because it is current and encyclopedic, but also because I teach Introductory Biology with Jim Gould.]

Krebs, J. R., and N. B. Davies. *An Introduction to Behavioural Ecol-*

ogy. Oxford: Blackwell, 2nd ed., 1987. [Here is the canonical textbook of this relatively new field that combines equal parts of natural history with analytical interpretation.]

Kricher, John C., and Gordon Morrison. *A Field Guide to Eastern Forests.* Boston, Mass.: Houghton Mifflin, 1988. [This is not really a field guide, but it is crammed with interpretive ecology, not only of the forests themselves but also of their inhabitants, and of the stages of development that an abandoned field goes through as it returns to forest. It is very engagingly written and deserves publication in a larger format.]

Lewontin, Richard C. *The Genetic Basis of Evolutionary Change.* New York: Columbia University Press, 1974. [This is the modern analog of Fisher's *Genetical Theory* . . . in Bates's list. A lot of the material is original and tough, but Lewontin is engaging and lucid.]

MacArthur, Robert H. *Geographical Ecology.* Princeton: Princeton University Press, 1984 reprint of 1972 edition. [This is now a classic manifesto of mathematical insights into the ecology of the real world. Large parts of it are accessible to a general audience.]

Mayr, Ernst. *Populations, Species, and Evolution.* Cambridge, Mass.: Harvard University Press, 1970 abridgment of 1963, *Animal Species and Evolution.* [This is Mayr's own development of the opus that Bates cites, thirty years later. Though technical, it is a literate and comprehensive discussion of the origin of species. It is worth reading both books to watch Mayr's ideas mature.]

Mitchell, A. *The Young Naturalist: An Introduction to Nature Studies.* London: Usborne (U.S. Agent: EDC Publishing, Tulsa, Okla.), 1982. [This is a seductive introduction to techniques and appropriate observations to make in virtually any setting. It is suitable for either an experienced child or a naive adult. The examples are English, but there are usually corresponding North American species.]

Purves, William K., and Gordon H. Orians. *Life: the Science of Biol-*

ogy. Sunderland, Mass. Sinauer, 2nd ed., 1986. [This is my choice for the most attractive and intelligible biology textbook for non-biologists.]

Roughgarden, J., R. M. May, and S. A. Levin, eds. *Perspectives in Ecological Theory*. Princeton: Princeton University Press, 1989. [This is a highly technical picture of the state of the art of analytical approaches to ecology. It is for specialists, and not even for the faint of heart among them. However, the best chapters are very clearly written. They show that ecologists have heeded Bates's pleas on pp. 69 and 275–277 for more outdoor experiments.]

Sheppard, P. M. *Natural Selection and Heredity*. London: Hutchinson & Co.: 2nd ed., 1975. [Despite its age, this is still my choice for the clearest précis of the facts and theories of evolution.]

Smith, John Maynard. *The Problems of Biology*. Oxford: Oxford University Press, 1986. [The introductory chapter of this book has the clearest exposition I have found of the molecular underpinnings of evolution. It shows that there has been some progress toward understanding the link between molecular genes and the material of evolution, though we are still far from the understanding of the further link to form and function that Bates hopes for on p. 225.]

Tattersall, Ian, et al. *Encyclopedia of Human Evolution & Prehistory*. New York: Garland Publishing, 1988. [This is the canonical reference for the facts and theories of our own evolution, and it updates Bates's brief reference on p. 19. The more hard evidence there is for patterns of human evolution, the more apparent it becomes that we make more artificial distinctions in interpreting our own history than we do in the rest of natural history.]

Thoreau, Henry David. *Excursions*. Gloucester, Mass.: Peter Smith, 1975 reprint of posthumous 1863 edition. [To my mind, this is the best contemporary collection of Thoreau's insights into nature. His *Journals* are also full of gems, though they are embed-

ded in a matrix of personal and societal introspection. A fair selection is available in an inexpensive paperback, *H. D. Thoreau,
The Natural History Essays*. Salt Lake City: Peregrine Smith,
1980.]

Tinbergen, N. *The Study of Instinct*. Oxford: Oxford University Press,
1989 reprint of 1951 edition. [At the time Bates wrote, this book
practically defined the field of ethology, the study of the mechanisms of animal behavior by interpretive observation. It also set
the stage for the developing field of behavioral ecology. Because
Tinbergen's work is based so firmly on perceptive observation, it
is still exemplary.]

van Andel, Tjeerd H. *New Views on an Old Planet: Continental Drift
and the History of the Earth*. Cambridge, England: The University Press, 1985. [This is an engaging and very clear account of
the current revolution in geological concepts, with special attention to implications for natural history. In particular, Bates's discussion of the geography of major patterns of evolution (pp. 194–
200) is affected, and there is more evidence for the importance of
unique events in earth's history than Bates assumes in pp. 46–
55.]

Wallace, Alfred Russel. *Natural Selection and Tropical Nature: Essays
on Descriptive and Theoretical Biology*. New York: Macmillan,
1895. [This is heavy going, but my favorite among Wallace's
many books on natural history.]

Watts, May T. *Reading the Landscape of America*. New York:
Macmillan, 1975. [Don't let the superficially cutesy-pie appearance of this book fool you. It is full of introductory insights into
geology and ecology, described with contagious enthusiasm. It
underscores the simple elegance of some important general principles of ecology.]

Wernert, Susan J., gen. ed. *North American Wildlife*. Pleasantville:
Reader's Digest, 1982. [Considering its range, this fat but handy
book has a surprisingly complete sample of species of plants and

animals that you are likely to encounter. It has nice pictures of each and brief notes on their natural history.]

White, Gilbert. *The Natural History of Selbourne*. New York: Penguin, 1977 edition of posthumous publication of 1789. [This is perhaps the first book written with a truly ecological perspective on the relationships among organisms in the natural world. Some of White's interpretations are dated and silly in retrospect, but those insights that depend directly on observation are as enlightening now as they were 200 years ago.]

Wilson, Edward O. *The Insect Societies*. Cambridge, Mass.: Harvard University Press, 1971. [Bates was a great fan of William Morton Wheeler. So is Wilson. Much of the modern extension of Wheeler's work on the natural history of ants has been done by Wilson. Like Wheeler, Wilson is a superb thinker and writer.]

Index

Cocoons, 83
Codification, 273
Coelenterates, 35; growth of, 76; reproduction of, 61
Collecting habit, 254
Commensalism, 125; in insects, 134
Community, definition of, 110; description of, 104; experimental study of, 121; as functioning unit, 123
Composites, 33, 207
Comstock, A. B., 289, 292
Conant, J. B., 4, 257, 269, 292
Concealing coloration, 209
Concepts, provisional, in science, 140
Conceptual schemes, 279
Conditioned reflex, 168
Conjugation of Paramecium, 63
Conservation, 242, 246
Continuity of life process, 57
Cooperation in nature, 108, 122
Coral snakes, 215
Corals, 35; symbiosis with algae, 127
Corynebacterium diphtheriae, 144
Crabs, 39; as hosts of Sacculina, 151
Cretaceous period, 48, 51
Cricetidae, 191
Crocodiles, 235
Cryptocotyle lingua, 150
Cryptostilis, 208
Crystallization of viruses, 26
Cultural vs. biological inheritance, 4; changes in, 5
Cultures, relations among, 111
Curators, 259
Curiosity as trait of scientists, 256
Cutright, P. R., 292
Cycles, of abundance, 170; of climate, 179; of culture, 5
Cynaelurus, as example of a genus, 18
Cytology, 7, 33

Darlington, Philip J., 194, 197, 292
Darwin, Charles, 13, 38, 43, 44, 57, 133, 134, 136, 159, 172, 187, 210, 221, 223, 224, 225, 226, 228, 236, 237, 242, 263, 292
Darwin, Erasmus, 223
Darwinians, 236
Daubenmire, R. F., 293
Davis, David, 178
Deer, young of, 77
Definition, difficulty of, 92, 140

Degeneracy of parasitism, 146, 151
Demography, 175
Dengue, 142
Descriptive method, 270
Deshayes, Gérard, 52
Desmids, 29
Despotism in nature, 164
Devonian period, 48, 50
Diatoms, 29
Dicotyledons, 32
Digestion, role of bacteria in, 130
Dinosaurs, 50
Diploid chromosome number, 62
Diphtheria, 143, 144
Disease, concept of, 143; relation to environment, 245
Dispersal mechanisms, 196, 197
Divisions of geological time, 46
Dobzhansky, Th., 172, 212, 293
Documentation in science, 285
Dodder, 144
Dogma, 280
Dogs, behavior comparisons with man, 163; distribution of, 201; as species, 14
Dolphins, 196
Dominants in community, 110
Dormancy, 182
Dormouse, 183
Drive, evolutionary, 235
Drosophila, 159, 174; chromosomes of, 68
Dry seasons, 180
Dunbar, C. O., 293

Earth, age of, 43
Earthworms, 37, 205; in soil formation, 134
Echinoderms, 38; in evolution, 88; metamorphosis of, 84
Ecology, 7, 90, 245
Edentates, 194
Education of naturalists, 256, 274
Eggs, 77; hibernation in, 182
Elements found in protoplasm, 102
Elephantiasis, 37
Elephants, 172, 234
Elton, Charles, 9, 91, 115, 118, 119, 120, 171, 174, 184, 293
Embryo, definition of, 76; development of, 85; environment of, 91; recapitulation in, 86